Lecture Notes in Control and Information Sciences

Edited by A. V. Balakrishnan and M. Thoma

Vol. 1: Distributed Parameter Systems: Modelling and Identification
Proceedings of the IFIP Working Conference, Rome, Italy, June 21–26, 1976
Edited by A. Ruberti
V, 458 pages. 1978

Vol. 2: New Trends in Systems Analysis
International Symposium, Versailles, December 13–17, 1976
Edited by A. Bensoussan and J. L. Lions
VII, 759 pages. 1977

Vol. 3: Differential Games and Applications
Proceedings of a Workshop, Enschede, Netherlands, March 16–25, 1977
Edited by P. Hagedorn, H. W. Knobloch, and G. J. Olsder
XII, 236 pages. 1977

Vol. 4: M. A. Crane, A. J. Lemoine
An Introduction to the Regenerative Method for Simulation Analysis
VII, 111 pages. 1977

Vol. 5: David J. Clements, Brian D. O. Anderson
Singular Optimal Control: The Linear Quadratic Problem
V, 93 pages. 1978

Vol. 6: Optimization Techniques
Proceedings of the 8th IFIP Conference on Optimization Techniques, Würzburg, September 5–9, 1977
Part 1
Edited by J. Stoer
XIII, 528 pages. 1978

Vol. 7: Optimization Techniques
Proceedings of the 8th IFIP Conference on Optimization Techniques, Würzburg, September 5–9, 1977
Part 2
Edited by J. Stoer
XIII, 512 pages. 1978

Vol. 8: R. F. Curtain, A. J. Pritchard
Infinite Dimensional Linear Systems Theory
VII, 298 pages. 1978

Vol. 9: Y. M. El-Fattah, C. Foulard
Learning Systems:
Decision, Simulation, and Control
VII, 119 pages. 1978

Vol. 10: J. M. Maciejowski
The Modelling of Systems with Small Observation Sets
VII, 241 pages. 1978

Vol. 11: Y. Sewaragi, T. Soeda, S. Omatu
Modelling, Estimation, and Their Applications for Distributed Parameter Systems
VI, 269 pages. 1978

Vol. 12: I. Postlethwaite, A. G. J. McFarlane
A Complex Variable Approach to the Analysis of Linear Multivariable Feedback Systems
IV, 177 pages. 1979

Vol. 13: E. D. Sontag
Polynomial Response Maps
VIII, 168 pages. 1979

Vol. 14: International Symposium on Systems Optimization and Analysis
Rocquentcourt, December 11–13, 1978;
IRIA LABORIA
Edited by A. Bensoussan and J. Lions
VIII, 332 pages. 1979

Vol. 15: Semi-Infinite Programming
Proceedings of a Workshop, Bad Honnef, August 30 – September 1, 1978
V, 180 pages. 1979

Vol. 16: Stochastic Control Theory and Stochastic Differential Systems
Proceedings of a Workshop of the „Sonderforschungsbereich 72 der Deutschen Forschungsgemeinschaft an der Universität Bonn"
which took place in January 1979 at Bad Honnef
VIII, 615 pages. 1979

Vol. 17: O. I. Franksen, P. Falster, F. J. Evans
Qualitative Aspects of Large Scale Systems
Developing Design Rules Using APL
XII, 119 pages. 1979

Vol. 18: Modelling and Optimization of Complex Systems
Proceedings of the IFIP-TC 7 Working Conference
Novosibirsk, USSR, 3–9 July, 1978
Edited by G. I. Marchuk
VI, 293 pages. 1979

Vol. 19: Global and Large Scale System Models
Proceedings of the Center for Advanced Studies (CAS)
International Summer Seminar
Dubrovnik, Yugoslavia, August 21–26, 1978
Edited by B. Lazarević
VIII, 232 pages. 1979

Vol. 20: B. Egardt
Stability of Adaptive Controllers
V, 158 pages. 1979

Vol. 21: Martin B. Zarrop
Optimal Experiment Design for Dynamic System Identification
X, 197 pages. 1979

For further listing of published volumes please turn over to inside of back cover.

Lecture Notes in Control and Information Sciences

Edited by A.V. Balakrishnan and M.Thoma

48

Yaakov Yavin

Feedback Strategies for Partially Observable Stochastic Systems

Springer-Verlag
Berlin Heidelberg GmbH 1983

Author
Dr. Yaakov Yavin
c/o NRIMS
CSIR
P.O. Box 395
Pretoria 0001 – South Africa

ISBN 978-3-540-12208-1 ISBN 978-3-540-39562-1 (eBook)
DOI 10.1007/978-3-540-39562-1

Originally published by Springer-Verlag Berlin Heidelberg New York in 1983

2061/3020-543210

PREFACE

The problems dealt with in this monograph are those of characterizing and computing optimal or suboptimal strategies for some classes of partially observable nonlinear stochastic systems. A method is developed by means of which sufficient conditions on weak optimal feedback strategies are established. These conditions are given by a set of coupled nonlinear partial integro-differential equations, in which the coupling constitutes a very complicated optimization problem. Numerical procedures are devised for computing weak suboptimal feedback strategies. Specific applications include strategies for the interception of moving or fixed targets by a pursuer, whose velocity law (strategy) is perturbed by a Gaussian white noise and where the observations available to it are either measurements of part of the state space, or different structures of noisy and interrupted observations, or noisy randomly sampled observations. Also treated are a rendezvous problem with partial observations; the control of a noisy nonlinear oscillator; and several problems of nonlinear filtering where an 'optimal' observer is to be designed.

The treatment is restricted in several ways. In the first place, only feedback strategies are considered; and although necessary conditions on optimal strategies are derived, efforts have been devoted mainly to the derivation of sufficient conditions on weak optimal strategies, and the computation of weak suboptimal strategies. Secondly, no existence theorems have been presented. Thus, the results presented here represent at best but a small brick is the edifice of modern control theory and its applications.

It is my pleasant duty to record here my sincere thanks to the

National Research Institute for Mathematical Sciences of the CSIR for encouraging the research on which most of the discussions is based.

Most of the work was partially supported by a grant from Control Data, which is also gratefully acknowledged.

I should like to thank the publishers of the International Journal of Systems Science, for permission to include in this monograph, substan= tial parts from references [40], [77], [78], [79], [80], [95] and [96], and similarly, the publishers of Computersand Mathematics with Applications and of Computer Methods in Applied Mechanics and Engineering, for permis= sion to include substantial parts from references [38] and [39] respec= tively.

Finally, I should like to thank Mrs M Russouw for her excellent typing of the manuscript.

Pretoria, September 1982 Yaakov Yavin

CONTENTS

CHAPTER 1

PRELIMINARIES

1.1 INTRODUCTION

This work deals with the problem of finding optimal or suboptimal feedback strategies v for systems whose dynamics are governed by a set of nonlinear stochastic differential equations, and where only incomplete observations of the state are available for v. The efforts here have been directed exclusively towards the derivation of procedures for com= puting optimal or suboptimal feedback strategies.

Under consideration are systems whose state process $X = \{X_t, t \geq 0\}$ is governed by an equation of the form

$$dx = [f(x) + F(x)v]dt + \sigma(x)dW + \int_{\mathbb{R}^m} c(x,u)q(dt,du) \tag{1.1}$$

$$t > 0, \quad x \in \mathbb{R}^m$$

with initial condition $X_0 = x_0$, where $f : \mathbb{R}^m \to \mathbb{R}^m$, $F : \mathbb{R}^m \to \mathbb{R}^{m\times d}$ ($\mathbb{R}^{m\times d}$ denotes the space of m×d matrices), $\sigma: \mathbb{R}^m \to \mathbb{R}^{m\times m}$ and $c : \mathbb{R}^m \times \mathbb{R}^m \to \mathbb{R}^m$ are given functions. $W \triangleq \{W(t) \triangleq (W_1(t),\ldots,W_m(t)), t \geq 0\}$ is an $\mathbb{R}^m$-valued standard Wiener process and q is a zero mean Poisson measure (see Gihman and Skorohod [1] or [2] for more details on q). $v : \mathbb{R}^p \to \mathcal{U}$, $\mathcal{U} \subseteq \mathbb{R}^d$, is a feedback strategy.

Let D_0 be an open and bounded set in $\mathbb{R}^m$ and let K be a closed set, $K \subset D_0$. Denote by τ_0 and τ the first exit times of X from D_0 and $D \triangleq D_0 - K$ respectively. The class of all functions $v : \mathbb{R}^p \to \mathcal{U}$ that are measurable and satisfy some other conditions to be specified later, is denoted by U. We refer to U as the class of *admissible strategies*. The strategy is to be so chosen as to maximize (or minimize) a criterion of the form

$$V(x_o;v) = E\{\int_0^{\tau_0} k(X_t,v)dt\} \tag{1.2}$$

or of the form

$$V(x_o;v) = \text{Prob}\{X_\tau \in K\}. \tag{1.3}$$

The following classes of information structures, available for v, are considered here.

(a) Only certain components of X_t can be observed : the observation process Y_t is given by

$$Y_t = (x_1(t),\ldots,x_p(t)),\ t \geq 0,\ p < m. \tag{1.4}$$

This kind of information structure includes the case where the observation process Y_t is determined by

$$dy = h(x)dt + \gamma(x)dB,\ t > 0,\ y \in \mathbb{R}^p \tag{1.5}$$

where $h : \mathbb{R}^m \to \mathbb{R}^p$ and $\gamma : \mathbb{R}^m \to \mathbb{R}^{p\times p}$ are given functions and $B \triangleq \{B(t) \triangleq (B_1(t),\ldots,B_p(t)),\ t \geq 0\}$ is an $\mathbb{R}^p$-valued standard Wiener process.

(b) Interrupted noisy observations: the observation process Y_t is given by

$$dy = \theta x dt + \gamma(x)dB,\quad t > 0,\quad y \in \mathbb{R}^m \quad \text{(here } p=m) \tag{1.6}$$

where $\gamma : \mathbb{R}^m \to \mathbb{R}^{m\times m}$ is a given function; B is an $\mathbb{R}^m$-valued standard Wiener process, and $\Theta = \{\theta(t),\ t \geq 0\}$ is a homogeneous jump Markov process with state space $S = \{0,1\}$.

(c) Interrupted observations: the observation process Y_t is given by

$$Y_t = \begin{cases} X_t = (x_1(t),\dots,x_m(t)) & \text{if } \theta(t) = 1 \\ \check{X}_t = (x_1(t),\dots,x_p(t)) & \text{if } \theta(t) = 0 \end{cases} \quad (1.7)$$

$p < m$, where Θ is described in (b). In this case we take the strategy v to be of the form

$$v(Y_t) = \theta(t)u(X_t) + (1 - \theta(t))\check{v}(\check{X}_t), \quad t \geq 0$$

and U denotes the class of all pairs $v = (u,\check{v})$, such that $u : \mathbb{R}^m \to \mathcal{U}_1$, $\check{v} : \mathbb{R}^p \to \mathcal{U}_0$, $\mathcal{U}_i \subset \mathbb{R}^d$, $i=0,1$, are measurable and satisfy some other conditions to be specified later.

(d) Random sampling of the observation : the observation process is given by

$$dy_i = x_i dN + \gamma_i(x)dB_i, \quad t > 0, \ i=1,\dots,m \quad \text{(here } p=m\text{)} \quad (1.8)$$

where $\gamma_i : \mathbb{R}^m \to \mathbb{R}$, $i=1,\dots,m$ are given functions; B is an $\mathbb{R}^m$-valued standard Wiener process, and $N \triangleq \{N(t), t \geq 0\}$ is a Poisson process.

A strategy $v^* \in U$ for which

$$V(x;v^*) \geq V(x;v) \quad \text{for any } v \in U \text{ and all } x \in D_0 \quad (1.9)$$

(where V is given by (1.2) or (1.3)) will here be called an *optimal strategy*.

In Chapter 2 the problem is considered of maximizing $V(\cdot;v)$ on U, where the information structure available to v is given by (1.4).

First, necessary conditions are derived for the cases:

(i) $U = \mathbb{R}^d$ and $v : \mathbb{R}^p \to \mathbb{R}^d$ are smooth functions;

(ii) $v : \mathbb{R}^p \to U$ are measurable functions and $U = \{x \in \mathbb{R}^d : |x_i| \le v_{0,i},$ $i=1,\ldots,d\}$, where $v_{0,i}$, $i=1,\ldots,d$ are given positive numbers.

In both cases these conditions amount to solving a pair of coupled non= linear partial integro-differential equations.

Next, the notion of weak optimal strategies is introduced, and suffi= cient conditions are derived, on weak optimal strategies of bang-bang type. As these conditions are given by a set of two coupled nonlinear partial integro-differential equations in which the coupling constitutes a very complicated optimization problem, a procedure is suggested for computing weak suboptimal strategies, and these are computed for three different examples.

In Chapter 3 problems are considered of maximizing $V(\cdot;v)$ on U, where $c(\cdot,\cdot) = 0$ and the information structure available to v is given by either (1.6) or (1.7) or (1.8). For each of the classes of information struc= ture, necessary conditions are derived on optimal strategies for cases (i) and (ii). When the information structure is described by (1.6) or (1.7), the necessary conditions amount to solving a set of four coupled nonlinear partial differential equations. This set of equations is solved numerically for a particular example. Also derived are sufficient con= ditions on weak optimal strategies which turn out to be given by a set of four coupled nonlinear partial differential equations in which the coupling constitutes a very complicated optimization problem. Again, as in Chapter 2, a procedure is suggested for computing weak suboptimal strategies, and these are computed for two different examples. For the case where the information structure available to v is given by (1.8), the results of Chapter 2 are applied to derive sufficient conditions on weak optimal strategies, the latter being computed for a particular example.

The problem of nonlinear filtering can be described as follows. $X_t, t \geq 0$, called the signal or the state of the system, is an $\mathbb{R}^m$-valued stochastic process, direct observation of which is not possible. The data related to X_t are provided by observation on an $\mathbb{R}^p$-valued process Y_t that is related to X_t either by the model (1.10),(1.5); or by (1.10), (1.6); or (1.10),(1.8); or by (1.10) and some jump process observation (see Segall [3] and Snyder [4], for example), or by some other combination. Using the least squares error criterion, the aim is to obtain the optimal estimate of X_t given the observed process $\{Y_s, s \leq t\}$. The optimal estimate is the conditional expectation $E[X_t \mid Y_s, s < t]$. It is known (see, for example, Fisher [5] and Kwakernaak [6]) that causal least squares state estimation for such systems in general requires real-time computation of the solutions of an infinite set of coupled stochastic differential equations in order to generate the estimate $\hat{X}_t = E[X_t \mid Y_s, 0 \leq s < t]$ of the state vector X_t. As a result, nonlinear filters (which compute $\{\hat{X}_t, t \geq 0\}$) have not yet become practical.

In Chapter 4 the problem of nonlinear estimation is approached by means of the concept of an observer for stochastic nonlinear systems. This approach may prove to have greater advantages for implementation than the approach by way of estimation theory.

Two classes of systems are considered.

(a) Let $X = \{X_t, t \geq 0\}$ and $Y = \{Y_t, t \geq 0\}$ be governed by the differential equations

$$dx = f(x)dt + \sigma(x)dW + \int_{\mathbb{R}^m} c(x,u)q(dt,du), t > 0, \ x \in \mathbb{R}^m, \quad (1.10)$$

$$dy = h(x)dt + \gamma(x)dB, \quad t > 0, \quad y \in \mathbb{R}^p. \quad (1.5)$$

For the system given by (1.10) and (1.5), a dynamic state estimator is chosen having the following form (see, for example, Tarn and Rasis [7]):

$$dz_i = f_i(z)dt + \sum_{j=1}^{p} g_{ij}(z)(dy_j - h_j(z)dt),\ t > 0,\ i=1,\dots,m \qquad (1.11)$$

where the functions g_{ij}, $i=1,\dots,m$, $j=1,\dots,p$ are yet to be determined. The matrix G

$$G(z) \triangleq \begin{pmatrix} g_{11}(z) & \cdots & g_{1p}(z) \\ \cdot & & \\ \cdot & & \\ \cdot & & \\ g_{m1}(z) & \cdots & g_{mp}(z) \end{pmatrix},\ z \in \mathbb{R}^m \qquad (1.12)$$

will here be called the *gain matrix* of the observer (1.11).

Denote by $Z = \{Z_t, t \geq 0\}$ the state of the observer (1.11) and define

$$V_T(\alpha;G) \triangleq E\,\Lambda\,\{t : 0 \leq t < \bar{\tau}_0,\ |X_t - Z_t| \leq \varepsilon\} \qquad (1.13)$$

$$V(\alpha;G) \triangleq \text{Prob}\{(X_{\bar{\tau}}, Z_{\bar{\tau}}) \in K_I\} \qquad (1.14)$$

where $\bar{\tau}_0$ is the first exit time of (X_t,Z_t) from $D_0 \times D_0$; $(X_0,Z_0) = \alpha = (x,z)$; $\varepsilon > 0$ is a given number; Λ is the Lebesgue measure on the real line; $K_I \triangleq \{\alpha = (x,z) : |x-z| \leq \varepsilon,\ \alpha \in D_0 \times D_0,\ \text{and}\ d(\alpha, \partial(D_0 \times D_0)) \geq \delta\}$ where $0 < \delta << 1$ is given and $\partial(D_0 \times D_0)$ denotes the boundary of $D_0 \times D_0$; and $\bar{\tau}$ is the first exit time of (X_t,Z_t) from $(D_0 \times D_0) - K_I$.

Let U denote the class of all gain matrices $G = G(z)$ of bang-bang type which satisfy some conditions to be specified later. The gain matrix G in (1.11) is so chosen as to maximize $V_T(\alpha;G)$ on U; or to maxi= mize $V(\alpha;G)$ on U. Using the results of Chapter 2, sufficient conditions on weak optimal gain matrices are derived and weak suboptimal gain matrices computed for several examples.

(b) Let $X = \{X_t, t \geq 0\}$ and $Y = \{Y_t, t \geq 0\}$ be governed by

$$dx = f(x)dt + \sigma(x)dW, \quad t > 0, \quad x \in \mathbb{R}^m \tag{1.15}$$

$$dy = c_0 dN, \quad t > 0, \quad y \in \mathbb{R}^p, \quad c_0 > 0. \tag{1.16}$$

Two cases are dealt with.

(b1) N_i, $i=1,\ldots,p$ are square integrable counting processes for all $t \geq 0$ ($N(t) \triangleq (N_1(t),\ldots,N_p(t))$), and $N(t) - \int_0^t \lambda(X_s)ds$ is an $\mathbb{R}^p$-valued martingale (on an appropriate probability space (Ω,F,P) with an increasing family $\{F_t\}_{t \geq 0}$ of sub-σ-fields of F), where $\lambda : \mathbb{R}^m \to \mathbb{R}^p$ is a given function.

(b2) $N(t) = \int_0^t \int_{\mathbb{R}^m} c(X_{s-},u)\nu(ds,du)$, where ν is a Poisson random measure on $[0,\infty) \times \mathbb{R}^m$ (see Gihman and Skorohod [2] or [8]), and $c : \mathbb{R}^m \times \mathbb{R}^m \to \mathbb{R}^p$ is a given function.

An observer of the form

$$dz = f(z)dt + G(y,z)(dy - c_0\lambda(z)dt), \; t > 0, \; z \in \mathbb{R}^m \tag{1.17}$$

is required in case (b1); and an observer of the form

$$dz = f(z)dt + G(y,z)(dy - \int_{\mathbb{R}^m} c(z,u)\pi(du)dt), \; t > 0, \; z \in \mathbb{R}^m \tag{1.18}$$

in case (b2), where $E\nu(t,A) = t\pi(A)$ for any Borel set A in $\mathbb{R}^m$. The matrix $G : \mathbb{R}^p \times \mathbb{R}^m \to \mathbb{R}^{m\times p}$ is called the *gain matrix* of the observer (1.17) (or (1.18)).

Let $\tilde{D}_0$ be an open and bounded set in $\mathbb{R}^{2m+p}$, and let τ_i be the first exit time of (X_t,Y_t,Z_t) from $\tilde{D}_0$ ($i=1$ corresponds to case (b1) and $i=2$ corresponds to case (b2)).

Denote by U the class of all gain matrices $G = G(y,z)$ of bang-bang type which satisfy some conditions to be specified later. In both cases, the gain matrix in (1.17) or (1.18) is so chosen as to maximize $V_T(\alpha;G)$ on U, where

$$V_T(\alpha;G) = E \Lambda \{t : 0 \le t < \tau_i, |X_t - Z_t| \le \varepsilon\} \tag{1.19}$$

and $(X_0,Y_0,Z_0) = \alpha = (x,y,z)$.

Using the results of Chapter 2, sufficient conditions on weak optimal gain matrices are derived for both cases, and weak suboptimal gain matrices computed for several examples.

The numerical results presented in this work for the examples which have been numerically solved, suggest that the weak suboptimal strategies of Chapters 2-3, and the weak suboptimal gain matrices of Chapter 4, are in many cases actually optimal strategies or optimal gain matrices, respectively.

1.2 A SHORT REVIEW

Optimal stochastic control problems have been extensively studied during the last two decades (see, for example, Fleming [9], Wonham [10], Fleming and Rishel [11], Krylov [12] and Gihman and Skorohod [13]). When X_t is completely observable (i.e. when, at each t, X_t is available to v), it is possible to choose strategies of the kind $v_t = v(X_t)$ and derive conditions of a dynamic programming type, as sufficient conditions for the maximization of $V(x_0;v)$, eqn (1.2) or (1.3) (see, for example, Fleming and Rishel [11] and Yavin and Jordaan [14]); which procedure consequently yields the maximization of $V(x_0;v)$.

The problem of the optimal control of partially observable diffusions (in this work the case where $c(\cdot,\cdot)=0$ in (1.1)), where the information structure available to v is given by (1.4), has been treated by Fleming [15]. There an existence theorem is stated in the class of bounded measurable time-dependent feedback strategies, and necessary conditions are derived on optimal feedback strategies, in terms of conditional ex= pectations. The problem was pursued further on in Ahmed and Teo [16,17]. Reed and Teo [18] considered certain problems related to the optimal feedback control of systems governed by a general Ito stochastic differen= tial equation that allows for jumps in the diffusion process (for example, the case where $c(\cdot,\cdot)\neq 0$ in (1.1)) and where the information structure is given by (1.4). It is shown there that these stochastic optimal feed= back control problems can be converted into problems of the optimal con= trol of deterministic systems described by a parabolic integro-partial-differential equation.

In Friedman and Yavin [19], a class of jump diffusion processes is dealt with and the admissible strategies $v_t = v(t,Y_t)$(where Y_t is given by (1.4)) are assumed to satisfy the growth condition and a local Lipschitz condition. By applying the calculus of variations, necessary conditions on optimal strategies are derived, these conditions being given by a pair of coupled nonlinear partial integro-differential equa= tions.

In Davis and Varaiya [20], Davis [21] and Elliott and Varaiya [22] the optimal control is discussed of partially observed systems whose dynamics are described by a system of stochastic functional differential equations. In Davis and Varaiya [20], controls are considered based on three types of information patterns: partial and complete observation of the past, and observation of the current state. In each case a principle of optimality is proved, and criteria are established for op= timality of dynamic programming type. Further work on controls that are allowed to depend on the complete past, was done in Davis [21]. In Elliott and Varaiya [22], where only some of the components of X_t are ob=

served, it is shown that a control (strategy) v^* adapted to the partially observed σ-fields, is optimal if it minimizes the conditional expectations of a Hamiltonian, when these expectations are taken with respect to the measure induced by any other control. Systems whose dynamics are governed by nonlinear stochastic functional differential equations have also been treated by Christopeit [23,24]. In [23] Christopeit considers the exis= tence of optimal controls in a class of asmissible controls v, where v is a function whose value at time t may depend at most on specified in= formation about the past of X up to time t. In [24], Christopeit deals with the problem of the existence of optimal controls when observations can be taken only at certain discrete times.

For nonlinear stochastic systems, necessary conditions on optimal controls, in the form of versions of stochastic maximum principles, have been derived in Kushner [25], Haussmann [26,27], Kwakernaak [28] and Arkin and Saksonov [29] (and the references cited there).

Further work on the control of partially observable stochastic sys= tems, from a point of view different from that of the rest of the works mentioned here, can be found in Fleming [30,31]. A rather complete sur= vey of control with incomplete information is given in Ahmed [32], and material covering the general mathematical theory of controlled stochastic processes is given in Gihman and Skorohod [13].

However, according to the results obtained in [20] - [31] it appears to be difficult to derive feasible procedures for the implementation of the various optimal control laws dealt with there.

Problems of optimal control for systems with jump Markov disturbances have been considered by several authors during the last two decades (see, for example, Wonham [10] , Krasovskii and Lidskii [33], Rishel [34],

Sworder [35] and Olsder and Suri [36]). In these references the process (X,θ) (where θ denotes the jump Markov process) is completely observable, and the admissible control laws are of the form $v_t = v(X_t, \theta(t))$. These make it possible to find conditions of dynamic programming type on opti= mal control laws.

In the case where the information structure is given by (1.6) (or (1.7)) the admissible strategies are of the form $v_t = v(Y_t)$. This ex= cludes the possibility of deriving implementable conditions of dynamic programming type for these cases. Even in the case of linear systems the problem of estimating X_t from $\{Y_s, 0 \leq s \leq t\}$, where Y is given by (1.6), leads to an infinite-dimensional filter (Sawaragi et al. [37]). This excludes the possibility of applying control laws of the form $v_t = v(\hat{X}_t, \hat{\theta}(t))$, where $\hat{X}_t = E[X_t \mid Y_s, 0 \leq s < t]$ and $\hat{\theta}(t) = E[\theta(t) \mid Y_s, 0 \leq s < t]$. A similar situation arises when the information structure is given by (1.7).

In Yavin and Venter [38] and Yavin [39], the case is considered where the information structure is given by (1.6). In Yavin and Venter [38] necessary conditions on smooth optimal strategies are derived and optimal strategies computed for an example. Yavin [39] derived sufficient con= ditions on weak optimal strategies of bang-bang type, and computed weak suboptimal strategies. In Yavin [40] the case is considered where the information structure is given by (1.7). Sufficient conditions on weak optimal strategies of bang-bang type are derived, and weak suboptimal strategies computed.

In engineering, the problem of estimating the state variables of a dynamical system, given observations of the output variables, is of fundamental importance.

A vast number of papers have dealt with linear estimation theory, which is the natural basis for nonlinear estimation. A fairly compre= hensive survey of linear filtering theory, and which includes more than 390 references, is that of Kailath [41]. The concept of the state and its use in estimation was introduced by Kalman. Kalman [42], and Kalman and Bucy [43] presented linear recursive equations with the related non= linear Riccati equations for the least squares estimate of X_t. The con= tinuous-time filter derived there is called the Kalman or Kalman-Bucy filter.

Extensive work on nonlinear filtering on stochastic continuous-time systems has been done, various approaches being used. For more details see, for example, Stratonovich [44], Kushner [45,46], Wonham [47], Bucy [48], Fisher [5], Bucy and Joseph [49], Jazwinski [50], Frost and Kailath [51] and McGarty [52]. In these works $\{X_t, t \in [0,T]\}$ is a diffusion process or a jump diffusion process (Fisher [5], McGarty [52]) and $\{Y_t, t \in [0,T]\}$ (the observation process) is a continuous process as given by (1.5).

The nonlinear filtering problem has also been treated for other mo= dels related to the properties of X_t, the state of the systems, and the measurements process Y_t, these models differing from, or being more general than those described in the references mentioned before. See, for example, Fujisaki et al. [53], Snyder [54], Clements and Anderson [55], Kwakernaak [6], Liptser and Shiryayev [56], Björk [57] and Rishel [58].

In most of the works mentioned, recursive formulae were obtained for updating the least-squares estimate $E[X_t \mid Y_s, 0 \leq s < t]$. It was found, however, that in general (exceptions are, for example, cases where X_t is a finite state process, Clements and Anderson [55], Liptser and

Siryayev [56], or some other cases, Björk [57], Rishel [58])the formulae involve all the conditional moments, so that an infinite set of coupled stochastic differential equations is formed.

There is still no consensus as to a satisfactory way of 'truncating' this set of equations. Furthermore, no other computationally satisfac= tory approaches seem as yet to have been made towards solving these sets of equations directly, even approximately. Consequently, depending on the assumptions stated, which are almost without exception heuristic, all practical algorithms are more or less approximate, having a somewhat poor basis in the strict mathematical sense. The best-known practical algorithms are linearized and extended Kalman filters (see Jazwinski [50]), and these have been used most frequently (see, for example, Mehra [59], Athans [60], McGarty [61], Dressler and Tabak [62], and Bucy and Joseph [49]).

The problem of estimating a signal X from jump process observations, is of fundamental importance in optical communications, nuclear medicine, estimation of traffic flow and some other fields. See, for example, Snyder [54,4], Snyder and Rhodes [63], Davidson and Carlson [64], Baras et al. [65] and Bagchi and Van Maarseveen [66]. A fairly comprehensive treatment of this problem is given in Snyder [4].

Extensive work on nonlinear filtering, using jump process observa= tions, has been done by applying martingale theory. See, for example, Van Schuppen [67,68], Segall [3], Segall and Kailath [69], Segall et al. [70], Vaca and Snyder [71,72], Gertner [73], Liptser and Shiryayev [56] and Boel and Beneš [74]. In these papers, as well as others, recursive formulae were obtained for updating the least squares optimal estimate $E[X_t \mid Y_s, 0 \le s < t]$, where Y is the observed jump process. It was found, however, that in general these formulae are merely a representation of the estimator, and not an explicit solution of the filtering problem (cases where these formulae reduce to explicit finite-dimensional recur= sive filters are given, for example, in Segall et al. [70], Wan and Davis [75], Davis and Andreadakis [76], and in Liptser and Shiryayev [56]). Boel and Beneš [74] derive explicit, though in general infinite-dimensional, recursive filters for three cases. Linearized filters are given in Snyder [4].

Summarizing, the usefulness of nonlinear filters (using continuous-in-time observations, or jump process observations) is in general limited by the complexity and number of on-line operations required. As a result, nonlinear filters have not yet become practical.

An alternative approach to nonlinear filtering is described in Tarn and Rasis [7], where observers for nonlinear stochastic systems are constructed based on a Lyapunov-like method. In Yavin and Friedman [77] and Yavin [78,79,80], observers for nonlinear stochastic systems are constructed by solving optimal control problems under partial observa= tion.

In this work, a method is developed by means of which for some classes of control problems under incomplete information,sufficient conditions on weak optimal feedback strategies are derived. As these conditions are given by a set of coupled nonlinear partial integro-differential equations in which the coupling constitutes a very complicated optimiza= tion problem, procedures are suggested for computing weak suboptimal feedback strategies, which is then done for a variety of examples.

CHAPTER 2

BANG-BANG PARTIALLY OBSERVABLE FEEDBACK STRATEGIES

2.1 INTRODUCTION

Let a nonlinear stochastic system be given by

$$dx = [f(x) + F(x)v(y)]dt + \sigma(x)dW + \int_{\mathbb{R}^m} c(x,u)q(dt,du)$$
$$t > 0,\ x \in \mathbb{R}^m \tag{2.1}$$

where y is a vector consisting of certain components of x, say

$$y = (x_1,\ldots,x_p),\ p < m; \tag{2.2}$$

$f : \mathbb{R}^m \to \mathbb{R}^m$, $F : \mathbb{R}^m \to \mathbb{R}^{m\times d}$, $\sigma : \mathbb{R}^m \to \mathbb{R}^{m\times m}$ and $c : \mathbb{R}^m \times \mathbb{R}^m \to \mathbb{R}^m$ are given functions; $v : \mathbb{R}^p \to \mathcal{U}$, $\mathcal{U} \subseteq \mathbb{R}^d$, is a feedback strategy; $W = \{W(t) \triangleq (W_1(t),\ldots,W_m(t)),\ t \geq 0\}$ is an $\mathbb{R}^m$-valued standard Wiener process; q is a zero-mean Poisson random measure on $[0,\infty) \times \mathbb{R}^m$, i.e.

$$q(t,A) = \nu(t,A) - t\pi(A),\ t \geq 0,\ A \in \mathcal{B}(\mathbb{R}^m) \tag{2.3}$$

where $\mathcal{B}(\mathbb{R}^m)$ denotes the m-dimensional Borel σ-algebra, and $\{\nu(t,A),\ t \geq 0\}$, $A \in \mathcal{B}(\mathbb{R}^m)$, is a Poisson process with

$$E\nu(t,A) = t\pi(A),\quad t \geq 0,\quad A \in \mathcal{B}(\mathbb{R}^m). \tag{2.4}$$

For more details on $\{\nu(t,A),\ t \geq 0,\ A \in \mathcal{B}(\mathbb{R}^m)\}$ see Gihman and Skorohod [1]. It is assumed here that $\pi(A) = \rho P_J(A)$, $A \in \mathcal{B}(\mathbb{R}^m)$, where P_J is a probability measure on $\mathcal{B}(\mathbb{R}^m)$ and ρ is a given positive number, and it is further assumed that the processes W and $\{q(t,A),\ t \geq 0\}$ are mutually independent for any $A \in \mathcal{B}(\mathbb{R}^m)$.

Denote by $\zeta_x^v = \{\zeta_x^v(t),\ t \geq 0\}$ the solution (in a sense that will be specified later) to eqn (2.1) such that $\zeta_x^v(0) = x$. Let D_o be an open and bounded set in $\mathbb{R}^m$ and let K be a closed set, $K \subset D_o$. Denote $D \triangleq D_o - K$ and define

$$\tau_o(x;v) \triangleq \begin{cases} \inf\{t : \zeta_x^v(t) \notin D_o \text{ when } \zeta_x^v(0) = x \in D_o\} \\ 0 \qquad \text{if } \zeta_x^v(0) \notin D_o \\ \infty \qquad \text{if } \zeta_x^v(t) \in D_o \text{ for all } t \geq 0 \end{cases} \tag{2.5}$$

and

$$\tau(x;v) \triangleq \begin{cases} \inf\{t : \zeta_x^v(t) \notin D \text{ when } \zeta_x^v(0) = x \in D\} \\ 0 \qquad \text{if } \zeta_x^v(0) \notin D \\ \infty \qquad \text{if } \zeta_x^v(t) \in D \text{ for all } t \geq 0 \end{cases} \tag{2.6}$$

where v is an admissible strategy.

Henceforward it is assumed in this work that $D_o = D_{oy} \times D_{o\tilde{x}}$ where $\tilde{x} = (x_{p+1},\dots,x_m)$, i.e. $x = (y,\tilde{x}) \in D_o$ iff $y \in D_{oy}$ and $\tilde{x} \in D_{o\tilde{x}}$.

2.2 NECESSARY CONDITIONS ON OPTIMAL STRATEGIES : $\mathcal{U} = \mathbb{R}^d$

It is assumed here that f_i and F_{ij}, i=1,...,m, j=1,...,d, are bounded and continuously differentiable on $\mathbb{R}^m$ and that σ_{ij}, i,j=1,...,m are bounded and twice continuously differentiable on $\mathbb{R}^m$. Also, it is assumed that c_i, i=1,...,m and P_J satisfy the following conditions:

(i) $\int_{\mathbb{R}^m} c_i^2(x,u)P_J(du) \leq \ell_o(1+|x|^2)$ for all $x \in \mathbb{R}^m$, i=1,...,m (2.7)
for some $0 < \ell_o < \infty$;

(ii) there is a constant a_o such that

$$\int_{\mathbb{R}^m} |c_i(x,u) - c_i(x',u)|^2 P_J(du) \leq a_o\ |x-x'|^2 \tag{2.8}$$

for all $x,x' \in \mathbb{R}^m$, i=1,...,m;

(iii) $\int_{\mathbb{R}^m} c_i(x,u)P_J(du) = 0$ for all $x \in \mathbb{R}^m$, $i=1,\dots,m$. (2.9)

(Condition (iii) is stated for the sake of convenience only.)

Let U denote the class of all strategies $v = v(y)$ such that

(a) there is an ℓ_0 $(0 < \ell_0 < \infty)$, for which

$$|v(y)|^2 \le \ell_0(1 + |y|^2) \text{ for all } y \in \mathbb{R}^p; \tag{2.10}$$

(b) v is continuously differentiable on $\mathbb{R}^p$.

(c) $\sup_{x \in D_0} E_x\tau_0(x;v) < \infty$ $(E_x \triangleq E[\cdot|\zeta_x^v(0) = x])$.

Here the following notations are used:

$$|x|^2 = \sum_{i=1}^m x_i^2 \qquad |v(y)|^2 = \sum_{i=1}^d v_i^2(y) \tag{2.11}$$

Under the assumptions on f,F,σ, and c it follows (Gihman and Skoro= hod [1]) that for a given strategy $v \in U$, (2.1) has a unique solution $\zeta_x^v = \{\zeta_x^v(t),\ t \ge 0\}$ with right continuous sample paths. Furthermore, ζ_x^v is a strong Markov process (Dynkin [81]) on a probability space denoted by (Ω,F,P_x).

Let $k : \mathbb{R}^m \to \mathbb{R}$, be a given bounded and continuous function. We use the notation

$$k(x,v) \triangleq k(x) - \sum_{i=1}^d \lambda_i\, v_i^2(y)\ , \quad x \in D_0 \tag{2.12}$$

where λ_i, $i=1,\dots,d$ are given positive numbers. Define the following functional:

$$V(x;v) \triangleq E_x \int_0^{\tau_0(x;v)} k(\zeta_x^v(t),v(\eta_x^v(t)))dt \tag{2.13}$$

where E_x denotes the expectation operation with respect to P_x and

$$\eta_x^v(t) \triangleq (\zeta_{x,1}^v(t),\dots,\zeta_{x,p}^v(t)),\ t \ge 0. \tag{2.14}$$

In this section, necessary conditions are derived on a strategy $v^* \in U$ for which

$$V(x;v^*) \geq V(x;v) \text{ for any } v \in U \text{ and all } x \in D_0. \quad (2.15)$$

Let $\mathcal{D}_0$ denote the class of all functions $V = V(x)$ such that: V is continuous on the closure $\bar{D}_0$ of D_0, and twice continuously differentiable on D_0; for any $v \in U$, $\mathcal{L}(v)V \in L_2(D_0)$, where

$$\begin{aligned} \mathcal{L}(v)V(x) = {} & \sum_{i=1}^{m} [f_i(x) + \sum_{j=1}^{d} F_{ij}(x)v_j(y)]\partial V(x)/\partial x_i \\ & + (\tfrac{1}{2}) \sum_{i,j=1}^{m} (\sigma(x)\sigma'(x))_{ij}\, \partial^2 V(x)/\partial x_i \partial x_j \\ & + \rho \int_{\mathbb{R}^m} [V(x + c(x,u)) - V(x)]P_J(du), \; x \in D_0 \end{aligned} \quad (2.16)$$

$\mathcal{L}(v)$ is the infinitesimal generator of the family of Markov processes $\{(\zeta_x^v, P_x),\ x \in D_0\}$.

Given $v \in U$, let $V \in \mathcal{D}_0$ be a solution to

$$\left.\begin{aligned} \mathcal{L}(v)V(x) &= -k(x,v), \quad x \in D_0 \\ V(x) &= 0 \quad x \notin D_0 \end{aligned}\right\} \quad (2.17)$$

then, by using the generalized Itô formula (Gihman and Skorohod [1]), it can be shown that

$$V(x) = V(x;v) = E_x \int_0^{\tau_0(x;v)} k(\zeta_x^v(t), v(\eta_x^v(t)))dt. \quad (2.18)$$

Throughout this Chapter it is assumed that for any $u \in \mathbb{R}^m$ the map= ping

$$\zeta = x + c(x,u) \quad (2.19)$$

maps $\mathbb{R}^m$ one-to-one onto itself and that the inverse mapping

$$x = C(\zeta,u) \tag{2.20}$$

is differentiable. Denote by $\Delta(\zeta,u)$ the Jacobian of the transformation (2.20).

Define, for $v \in U$,

$$\begin{aligned}\mathcal{L}^*(v)Q(x) \triangleq &- \sum_{i=1}^{m} \partial[(f_i(x) + \sum_{j=1}^{d} F_{ij}(x)v_j(y))Q(x)]/\partial x_i \\ &+ \tfrac{1}{2} \sum_{i,j=1}^{m} \partial^2[(\sigma(x)\sigma'(x))_{ij}Q(x)]/\partial x_i\, \partial x_j \\ &+ \rho \int_{\mathbb{R}^m} [Q(C(x,u))\Delta(x,u) - Q(x)]P_J(du)\end{aligned} \tag{2.21}$$

for any Q such that $\mathcal{L}^*(v)Q \in L_2(D_o)$, and let J denote the following functional:

$$J(v) \triangleq \int_{D_o} V(x;v)dx \ , \ v \in U. \tag{2.22}$$

The following theorem gives necessary conditions on v^*.

<u>*Theorem 2.1*</u>

Suppose there exists a strategy $v^* \in U$ such that

$$V(x;v^*) \geq V(x;v) \text{ for any } v \in U \text{ and all } x \in D_o. \tag{2.23}$$

Let $v^\alpha = v^* + \alpha\psi$ for all $\alpha \in [0,\alpha_o]$, $\alpha_o > 0$, $\psi \in U$, (note that $v^\alpha \in U$, $\alpha \in [0,\alpha_o]$). Assume:

(i) for each $\alpha \in [0,\alpha_o]$ there is a unique function $V^\alpha \in \mathcal{D}_o$ satisfying

$$\left.\begin{aligned}&\mathcal{L}(v^\alpha)V^\alpha(x) = -k(x,v^\alpha) \ , \quad x \in D_o \\ &V^\alpha(x) = 0 \quad , \quad x \notin D_o\end{aligned}\right\} \tag{2.24}$$

(ii) there is a function Q_o satisfying

$$\left.\begin{aligned} &\mathcal{L}^*(v^*)Q_o(x) = -1 \quad , \text{ a.e. in } D_o \\ &Q_o(x) = 0 \quad , \quad x \notin D_o, \end{aligned}\right\} \tag{2.25}$$

(iii) $\partial V^{\alpha}/\partial x_i$, i=1,...,m, converge weakly (in $L_2(D_o)$) as $\alpha \downarrow 0$ to $\partial V^o/\partial x_i$, i=1,...,m, respectively.

Then

$$v_j^*(y) = [2\lambda_j \int_{D_{o\tilde{x}}} Q_o(x)d\tilde{x}]^{-1} \sum_{i=1}^{m} \int_{D_{o\tilde{x}}} Q_o(x)F_{ij}(x)(\partial V(x;v^*)/\partial x_i)d\tilde{x} \tag{2.26}$$

$j=1,...,d \quad , y \in D_{oy}.$

Proof

For each $\alpha \in [0,\alpha_o]$, let $V^{\alpha} \in \mathcal{D}_o$ satisfy (2.24); then

$$\mathcal{L}(v^*)(V^{\alpha}-V^*)+(\mathcal{L}(v^{\alpha})-\mathcal{L}(v^*))V^{\alpha} + k(\cdot,v^{\alpha})-k(\cdot,v^*) = 0 \tag{2.27}$$

where $V^* = V(\cdot;v^*)$.

If now the operation $\int_{D_o} Q_o$ is applied to both sides of (2.27), and use is made of (2.25) and (2.22), then (2.27) and (2.23) yield

$$\begin{aligned} J(v^{\alpha}) - J(v^*) = \alpha \sum_{j=1}^{d} \int_{D_{oy}} \psi_j(y) \int_{D_{o\tilde{x}}} Q_o(x) \, [\sum_{i=1}^{m} F_{ij}(x)\partial V^{\alpha}(x)/\partial x_i \\ - 2\lambda_j v_j^*(y)]d\tilde{x}dy \\ - \alpha^2 \sum_{j=1}^{d} \lambda_j \int_{D_o} Q_o(x) \psi_j^2(y)dx \le 0. \end{aligned} \tag{2.28}$$

Assuming that condition (iii) of Theorem 2.1 is satisfied, it follows that

$$\lim_{\alpha \to 0} (J(v^{\alpha}) - J(v^*))/\alpha$$

$$= \sum_{j=1}^{d} \int_{D_{oy}} \psi_j(y) \int_{D_{o\tilde{x}}} Q_o(x) \left[\sum_{i=1}^{m} F_{ij}(x)\partial V(x;v^*)/\partial x_i - 2\lambda_j v_j^*(y) \right] d\tilde{x}dy \tag{2.29}$$

$$\leq 0 \quad \text{for any } \psi \in U.$$

Hence v^* is given by (2.26). □

Thus if one assumes that a strategy $v^* \in U$ for which (2.23) is satis= fied exists, and that all the conditions stated in Theorem 2.1 are satis= fied, then in order to implement such a strategy, the following system of equations has to be solved:

$$\mathcal{L}(v)V(x) = -k(x,v) \quad , x \in D_o \tag{2.30}$$

$$\mathcal{L}^*(v)Q(x) = -1 \quad , \quad \text{a.e. in } D_o \tag{2.31}$$

$$V(x) = Q(x) = 0 \quad , x \notin D_o \tag{2.32}$$

where

$$v_j(y) = [2\lambda_j \int_{D_{o\tilde{x}}} Q(x)d\tilde{x}]^{-1} \sum_{i=1}^{m} \int_{D_{o\tilde{x}}} Q(x)F_{ij}(x)(\partial V(x)/\partial x_i)d\tilde{x} \tag{2.33}$$

$$j=1,\ldots,d \quad , \quad y \in D_{oy}.$$

Remark 2.1

Suppose there exists a strategy $v^o \in U$ such that

$$J(v^o) \geq J(v) \text{ for any } v \in U. \tag{2.34}$$

Let $v^{\alpha} = v^o + \alpha\psi$, for all $\alpha \in [0,\alpha_o]$, $\alpha_o > 0$, $\psi \in U$, and assume that conditions (i), (ii) and (iii) of Theorem 2.1 are satisfied. Then, by applying the same proof as in Theorem 2.1, it follows that v^o is given by

$$v_j^o(y) = [2\lambda_j \int_{D_{o\tilde{x}}} Q_o(x)d\tilde{x}]^{-1} \sum_{i=1}^{m} \int_{D_{o\tilde{x}}} Q_o(x)F_{ij}(x)(\partial V(x;v^o)/\partial x_i)d\tilde{x} \tag{2.35}$$

$$j=1,\ldots,d \quad , \quad y \in D_{oy},$$

and in order to implement v^o, the set of equations (2.30)-(2.33) has to be solved.

Remark 2.2

Let $v^* \in U$ and $v^o \in U$ satisfy (2.23) and (2.34) respectively. Then it can be shown that $V(x;v^*) = V(x;v^o)$ a.e. in D_o. Hence a strategy $v^o \in U$ that maximizes J on U, whenever it exists, can be interpreted as a solution, in some weak sense, to the problem: Find a strategy $v^* \in U$ such that

$$V(x;v^*) \geq V(x;v) \text{ for any } v \in U \text{ and all } x \in D_o. \tag{2.36}$$

2.3 NECESSARY CONDITIONS ON OPTIMAL STRATEGIES: $\mathcal{U}$ IS BOUNDED

It is assumed here that f, F,σ, c and P_J satisfy all the conditions stated in Section 2.2. Let

$$\mathcal{U} = \{x \in \mathbb{R}^d : |x_i| \leq v_{o,i}, \; i=1,\ldots,d\} \tag{2.37}$$

where $v_{o,i}$, i=1,...,d are given positive numbers. Denote by $\tilde{U}$ the class of all strategies $v = \{v(y) : y \in \mathbb{R}^p\}$ such that $v : \mathbb{R}^p \to \mathcal{U}$ is measurable. In order to have as strategies functions that are smooth enough, the following approach is adopted. Define for $v \in \tilde{U}$

$$v_a(y) \triangleq \int_{\mathbb{R}^p} v(y')\theta_a(y - y')dy' \tag{2.38}$$

where, for $0 < a \ll 1$,

$$\theta_a(y) \triangleq \begin{cases} \exp((|y/a|^2-1)^{-1})/h_a & \text{for } |y| = |(x_1,\ldots,x_p)| = [\sum_{i=1}^{p} x_i^2]^{\frac{1}{2}} < a \\ 0 & \text{for } |y| \geq a \end{cases} \quad (2.39)$$

where $h_a > 0$ is such that $\int_{\mathbb{R}^p} \theta_a(y)dy = 1$. The function v_a is continuously differentiable (see Yosida [82], for example) on $\mathbb{R}^p$ and $v_a(y) \in \mathcal{U}$ for all $y \in \mathbb{R}^p$. Henceforward in this section, instead of (2.1), the following equation is considered:

$$dx = [f(x) + F(x)v_a(y)]dt + \sigma(x)dW + \int_{\mathbb{R}^m} c(x,u)q(dt,du) \quad (2.40)$$

$$t > 0, \qquad x \in \mathbb{R}^m.$$

Under the assumptions on f, F, σ, c, P_J and v_a, (2.40) has a unique solution $\zeta_x^v = \{\zeta_x^v(t), t \geq 0\}$. Furthermore, ζ_x^v is a strong Markov process. We here use the notations ζ_x^v and $\tau_o(x;v)$ (rather than $\zeta_x^{v_a}$ and $\tau_o(x;v_a)$ respectively) since the strategy v uniquely determines the strategy v_a. We denote by U the following class of strategies $U \triangleq \{v \in \tilde{U} : \sup_{x \in D_o} E_x\tau_o(x;v)<\infty\}$ Define the class $\mathcal{D}_o$ in the same manner as in Section 2.2 but with respect to the class U as defined in this section.

Let $k : \mathbb{R}^m \to \mathbb{R}$, be a given bounded and continuous function. Define the following functional:

$$V(x;v) \triangleq E_x \int_0^{\tau_o(x;v)} k(\zeta_x^v(t))dt. \quad (2.41)$$

In this section necessary conditions are derived on a strategy $v^* \in U$ for which

$$V(x;v^*) \geq V(x;v) \text{ for any } v \in U \text{ and all } x \in D_o. \quad (2.42)$$

The following theorem gives necessary conditions on v^*.

Theorem 2.2

Suppose there exists a strategy $v^* \in U$ which satisfies (2.42). Assume

(i) for any strategy $v \in U$ there exists a unique solution $V(\cdot;v) \in \mathcal{D}_o$ to the equations

$$\mathcal{L}(v_a)V(x) = -k(x),\ x \in D_o;\quad V(x) = 0,\ x \notin D_o \tag{2.43}$$

where $\mathcal{L}(v)$ is defined by (2.16) and v_a is defined by (2.38)-(2.39);

(ii) there is a function Q_o satisfying

$$\mathcal{L}^*(v_a^*)Q_o(x) = -1 \text{ a.e. in } D_o;\quad Q_o(x) = 0,\ x \notin D_o \tag{2.44}$$

where $\mathcal{L}^*(v)$ is defined by (2.21);

(iii) there is at least one strategy $v^{(1)} \in U$ which satisfies

$$v_j(y) = v_{o,j} \text{ sign } \psi_j(y;v),\ j=1,\dots,d, \text{ a.e. in } D_{oy} \tag{2.45}$$

where

$$\psi_j(y;v) \triangleq \sum_{i=1}^{m} \int_{D_{oy}} \theta_a(y-y') \int_{D_{o\tilde{x}}} Q_o(y',\tilde{x})F_{ij}(y',\tilde{x})(\partial V(y',\tilde{x};v)/\partial x_i)d\tilde{x}dy' \tag{2.46}$$

$j=1,\dots,d,\quad y \in D_{oy}$.

Then

$$v_j^*(y) = v_{o,j} \text{ sign } \psi_j(y;v^*),\ j=1,\dots,d, \text{ a.e. in } D_{oy}. \tag{2.47}$$

(Without loss of generality, we here take $v(y) = 0$, $y \notin D_{oy}$, for any $v \in U$.)

Proof

From (i) it follows that, for any $v \in U$,

$$\mathcal{L}(v_a^*)(V(\cdot;v)-V(\cdot;v^*)) + (\mathcal{L}(v_a)-\mathcal{L}(v_a^*))V(\cdot;v) = 0. \tag{2.48}$$

Applying the operation $\int_{D_0} Q_0$ to both sides of (2.48); using (2.44), (2.22) and (2.42); and performing some manipulation; we obtain

$$J(v) - J(v^*) = \sum_{j=1}^{d} \int_{D_{oy}} (v_j(y)-v_j^*(y))\psi_j(y;v)dy \le 0. \tag{2.49}$$

Choose $v = v^{(1)} \in U$ such that

$$v_j^{(1)}(y) = v_{o,j}\ \mathrm{sign}\psi_j(y;v^{(1)}),\ j=1,\dots,d,\ \text{a.e. in } D_{oy}, \tag{2.50}$$

then

$$J(v^{(1)})-J(v^*) = \sum_{j=1}^{d} \int_{D_{oy}} \{v_{o,j}|\psi_j(y;v^{(1)})|-v_j^*(y)\psi_j(y;v^{(1)})\}dy \ge 0. \tag{2.51}$$

Denote by $|B|$ the Lebesgue measure of a measurable subset B of D_{oy}. Assume that for some j

$$v_j^*(y) \ne v_{o,j}\ \mathrm{sign}\ \psi(y;v^{(1)}) \text{ on a set } B \subset D_{oy},\ |B| > 0. \tag{2.52}$$

Then (2.51) and (2.52) imply

$$J(v^{(1)}) - J(v^*) > 0 \tag{2.53}$$

which contradicts (2.49). Hence

$$v_j^*(y) = v_{o,j}\ \mathrm{sign}\ \psi(y;v^{(1)}),\ j=1,\dots,d \quad \text{a.e. in } D_{oy}, \tag{2.54}$$

from which (2.47) follows. □

Remark 2.3

Theorem 2.2 remains valid if the assumption : $v^* \in U$ satisfies

$$J(v^*) \geq J(v) \text{ for any } v \in U \tag{2.55}$$

replaces the assumption implied by (2.42).

To summarize. If it is assumed that a strategy $v^* \in U$ exists, for which (2.42) or (2.55) is satisfied, and that conditions (i)-(iii) of Theorem 2.2 are satisfied, then, in order to implement such a strategy, the following set of equations has to be solved:

$$\mathcal{L}(v_a)V(x) = -k(x), \quad x \in D_0 \tag{2.56}$$

$$\mathcal{L}^*(v_a)Q(x) = -1 \quad , \quad \text{a.e. in } D_0 \tag{2.57}$$

$$V(x) = Q(x) = 0, \quad x \notin D_0 \tag{2.58}$$

$$v_j(y) = v_{0,j} \text{ sign } \psi_j(y;v), \; j=1,\dots d, \text{ a.e. in } D_{oy} \tag{2.59}$$

$$v_a(y) = \int_{D_{oy}} v(y')\theta_a(y - y')dy' \quad , \quad y \in D_{oy}. \tag{2.60}$$

2.4 FURTHER REMARKS ON THE NECESSARY CONDITIONS

The methods used in deriving Theorems 2.1 and 2.2 can be applied, with only slight changes, to the case where $V(x;v)$ is given by (1.3), or to some other cases, as for example the case treated in Friedman and Yavin [83].

The problem of the existence of optimal strategies $v^* \in U$, for the cases dealt with in Theorems 2.1 or 2.2, has not yet been studied. This situation is in sharp contrast to the considerable research efforts that have been devoted to an investigation of the existence of optimal strategies for time-dependent controlled diffusions.

Also, there is still a lack of knowledge about the existence and uniqueness of solutions to eqns (2.30)-(2.33) or eqns (2.56)-(2.60). For these reasons the applicability of Theorems 2.1 and 2.2 is restricted. It is, however, possible to find numerical approximations to the solutions of (2.30)-(2.33) (see Yavin and Friedman [77]). Such numerical solutions might be useful in practice. Also, some experience has been gained in solving numerically problems similar to that implied by (2.30)-(2.33), see for example Friedman and Yavin [83-84], and Yavin and Friedman [85].

2.5 WEAK OPTIMAL STRATEGIES

Consider the nonlinear stochastic system given by (2.1) and (2.2). It is assumed that f_i, $i=1,\dots,m$ are bounded and continuously differen= tiable on $\mathbb{R}^m$; that F_{ij}, $i=1,\dots,m$, $j=1,\dots,d$ are bounded and continuous on $\mathbb{R}^m$; that $(\sigma(x)\sigma'(x))_{ij}$, $i,j=1,\dots,m$ are bounded and twice continuously differentiable on $\mathbb{R}^m$; and that there exist constants $0 < \lambda_1 \le \lambda_2 < \infty$ such that $\lambda_1|\zeta|^2 \le (\zeta,\sigma(x)\sigma'(x)\zeta) \le \lambda_2|\zeta|^2$ for any $x,\zeta \in \mathbb{R}^m$.

Define, for a Borel set A in $\mathbb{R}^m$,

$$M(x,A) \triangleq P_J(u : c(x,u) \in A),\quad x \in \mathbb{R}^m \tag{2.61}$$

and assume that conditions (2.8) and (2.9) are satisfied by c and P_J. Then the operator $\mathcal{L}(v)$ (eqn (2.16)) is written as

$$\begin{aligned}\mathcal{L}(v)V(x) = &\sum_{i=1}^{m} [f_i(x) + \sum_{j=1}^{d} F_{ij}(x)v_j(y)]\partial V(x)/\partial x_i \\ &+ (\tfrac{1}{2}) \sum_{i,j=1}^{m} (\sigma(x)\sigma'(x))_{ij}\, \partial^2 V(x)/\partial x_i\partial x_j \\ &+ \rho\int_{\mathbb{R}^m} [V(x+\alpha) - V(x)]M(x,d\alpha),\quad V \in \mathcal{D}_0.\end{aligned} \tag{2.62}$$

Assume that

$\int_{\mathbb{R}^m}(|\alpha|^2/(1+|\alpha|^2))\phi(\alpha)M(x,d\alpha)$ and $\int_A(\alpha/(1+|\alpha|^2))M(x,d\alpha)$ are bounded and continuous (2.63)

on $\mathbb{R}^m$ for any Borel set A in $\mathbb{R}^m$, and all $\phi \in C_b(\mathbb{R}^m)$. ($C_b(\mathbb{R}^m)$ is the set of bounded continuous functions $f : \mathbb{R}^m \to \mathbb{R}$).

Denote by $\tilde{U}$ the class of all strategies $v = \{v(y) : y \in \mathbb{R}^p\}$ such that $v : \mathbb{R}^p \to \mathcal{U}$ is measurable and $\mathcal{U}$ is given by (2.37).

Given $x \in D_0$ and $v \in \tilde{U}$. Then under the assumptions on f, F, σ,c ,P_J and M, eqns (2.1)-(2.2) determine a stochastic process $\zeta_x^v = \{\zeta_x^v(t) = (\zeta_{x,1}^v(t),\ldots,\zeta_{x,m}^v(t)),\ t \geq 0\}$ such that:

(i) on a probability space (Ω,F,P_x^v), ζ_x^v is a weak solution to (2.1)-(2.2) in the sense that P_x^v is the unique solution to the martin= gale problem for $\mathcal{L}(v)$ (Stroock [86], Mahno [87], and Komatsu [88]);

(ii) $\zeta_x^v(0) = x$ P_x^v-almost surely;

(iii) the family $\{P_x^v : x \in D_0\}$ is strong Markov (Stroock [86]).

For more details on weak solutions to stochastic differential equa= tions, see also Stroock and Varadhan [89], Liptser and Shiryayev [56] and Beneš [90].

Let $U \triangleq \{v \in \tilde{U} : \sup_{x \in D} E_x^v\ \tau(x;v) < \infty\}$, (E_x^v denotes the expectation operation with respect to P_x^v).

We assume that for each $v \in U$, $\zeta_x^v(0)$ is an $\mathbb{R}^m$-valued random element with

$$P_x^v(\{\zeta_x^v(0) \in B\}) = \mu(B) \tag{2.64}$$

for all $B \in \mathcal{B}(\mathbb{R}^m)$, (where $\mathcal{B}(\mathbb{R}^m)$ denotes the σ-algebra of Borel sets of $\mathbb{R}^m$) and μ is a probability measure on $\mathcal{B}(\mathbb{R}^m)$ such that : $\mu(B) = 0$ if $B \cap \bar{D}_0 = \phi$ where ϕ is an empty set. Define for $v \in U$:

$$C(v) \triangleq \int_{D_0} P_x^v(\{\zeta_x^v(\tau(x;v)) \in K\})\mu(dx). \tag{2.65}$$

In words: $P_x^v(\{\zeta_x^v(\tau(x;v)) \in K\})$ is the probability (P_x^v) of the event $\{\zeta_x^v(t)$ enters the set K before any other subset of $D^c \mid \zeta_x^v(0) = x\}$ (D^c denotes the complement of D). Since $\zeta_x^v(0)$ is a random element, the functional $C(v)$ is introduced. Define the following probability measure:

$$P^v(A) \triangleq \int_{D_0} P_x^v(A)\mu(dx), \quad A \in F. \tag{2.66}$$

Then $C(v)$ is the probability (P^v) of the event $\{\zeta_x^v(t)$ enters the set K before any other subset of $D^c\}$.

The problem considered here is the following: find a strategy $v^* \in U$ such that

$$C(v^*) \geq C(v) \text{ for any } v \in U. \tag{2.67}$$

A strategy $v^* \in U$ for which (2.67) is satisfied will here be called an *optimal strategy*.

Denote

$$V(x;v) \triangleq P_x^v(\{\zeta_x^v(\tau(x;v)) \in K\}) \tag{2.68}$$

and define

$$L(v) \triangleq \int_{D_0} (1 - V(x;v))^2 dx \ , \ v \in U. \tag{2.69}$$

Let $v^* \in U$ and $v^o \in U$ satisfy $V(x;v^*) \geq V(x;v)$ for any $v \in U$ and all $x \in D_0$, and let $L(v^o) \leq L(v)$ for any $v \in U$. Then it can be shown that $V(x;v^*) = V(x;v^o)$ a.e. in D_0, and consequently $C(v^*) = C(v^o) \geq C(v)$ for any $v \in U$. Hence a strategy $v^o \in U$ that minimizes $L(v)$ on U, when= ever such a strategy exists, can be interpreted as an optimal strategy in some weak sense. A strategy $v^o \in U$ for which

$$L(v^0) \leq L(v) \text{ for any } v \in U \qquad (2.70)$$

will here be called a *weak optimal strategy*.

Later in this work, weak optimal strategies are defined for cost functions other than those given by (2.65) and (2.68)-(2.69). The method described in the next section, for deriving sufficient conditions on weak optimal strategies, remains the same when other cost functions are considered.

2.6 SUFFICIENT CONDITIONS ON WEAK OPTIMAL STRATEGIES

It is assumed here that f,F,σ,c,P_J,M and U satisfy all the condi= tions stated in Section 2.5.

Let $\mathcal{D}_0$ denote the class of all functions $V : \mathbb{R}^m \to \mathbb{R}$ such that: V is con= tinuous on the closure $\bar{D}_0$ of D_0 and twice continuously differentiable on D; for any $v \in U$, $\mathcal{L}(v)V \in L_2(D_0)$, where the class U is defined in Section 2.5.

Given $v \in U$, let $V \in \mathcal{D}_0$ be a solution to

$$\mathcal{L}(v)V(x) = 0, \quad x \in D \qquad (D = D_0 - K) \qquad (2.71)$$

$$V(x) = 1, \quad x \in K; \quad V(x) = 0, \quad x \notin D_0 \qquad (2.72)$$

where $\mathcal{L}(v)$ is given by (2.16). Then it can be shown (see, for example, Yavin and Reuter [91]) that

$$V(x) = V(x;v) = P_x^v(\{\zeta_x^v(\tau(x;v)) \in K\}), \; x \in \bar{D}_0. \qquad (2.73)$$

In the sequel the following lemma will be used.

Lemma 2.1

Let H be a Hilbert space, A a subset of H. Let V_d be an element of H, $V_d \notin \bar{A}$. Then in order that an element $V^* \in A$ should satisfy

$$\|V_d - V^*\| \leq \|V_d - V\| \quad \text{for any } V \in A \tag{2.74}$$

it is sufficient that there should exist an element $\psi \in H$ such that

(a) $$< \psi, V^* > = \sup_{V \in A} < \psi, V > \tag{2.75}$$

(b) $$\psi = V_d - V^*. \tag{2.76}$$

Proof

From (2.75) and (2.76) it follows that

$$\begin{aligned} \|V_d - V^*\|^2 &\leq < \psi, V_d - V^* > = < \psi, V_d > - < \psi, V^* > \\ &\leq < \psi, V_d > - < \psi, V > = < \psi, V_d - V > \\ &\leq \|\psi\| \|V_d - V\| = \|V_d - V^*\| \|V_d - V\| \text{ , for any } V \in A. \end{aligned} \tag{2.77}$$

Using the fact that $V_d \notin \bar{A}$, inequalities (2.77) imply

$$\|V_d - V^*\| \leq \|V_d - V\| \quad \text{for any } V \in A.$$ □

We here take $H \triangleq L_2(D_0)$.

Let $V \in \mathcal{D}_0$, and define the following operators

$$\mathcal{L}V(x) \triangleq \mathcal{L}(v)V(x) - \sum_{i=1}^{m} \sum_{j=1}^{d} F_{ij}(x)v_j(y)\partial V(x)/\partial x_i \tag{2.78}$$

and

$$\mathcal{L}^*Q(x) \triangleq \mathcal{L}^*(v)Q(x) + \sum_{i=1}^{m} \sum_{j=1}^{d} \partial[F_{ij}(x)v_j(y)Q(x)]/\partial x_i \text{ ,} \tag{2.79}$$

for all functions Q such that $\mathcal{L}^*Q \in L_2(D_0)$, where $\mathcal{L}(v)$ and $\mathcal{L}^*(v)$ are given by (2.16) and (2.21) respectively.

Suppose that $V(\cdot;v) \in \mathcal{D}_0$ and Q_0 are such that

(i) $V(\cdot;v)$ satisfies (2.71)-(2.72) for a given $v \in U$;

(ii) $\mathcal{L}^*Q_o \in L_2(D_o)$ and $Q_o(x) = 0 \quad x \in K \cup D_o^c$.

Then

$$\mathcal{L}(v)V(x) = I_K(x)\ \rho \int_{\mathbb{R}^m} [V(x+c(x,u))-1]P_J(du) \quad \text{a.e. in } D_o \tag{2.80}$$

(where $I_K(x) = 1$ if $x \in K$ and $I_K(x) = 0$ if $x \notin K$) and

$$\int_{D_o} V(x;v)\mathcal{L}^*Q_o(x)dx = \int_{D_o} Q_o(x)\mathcal{L}V(x)dx \tag{2.81}$$

and by using (2.78) and (2.80), and also (ii), we obtain

$$\begin{aligned}\int_{D_o} V(x;v)\mathcal{L}^*Q_o(x)dx &= \int_{D_o} Q_o(x)\ [-\sum_{i=1}^{m}\sum_{j=1}^{d} F_{ij}(x)v_j(y)\partial V(x;v)/\partial x_i \\ &\quad + I_K(x)\ \rho \int_{\mathbb{R}^m} [V(x+c(x,u)) - 1]P_J(du)]dx \\ &= -\sum_{j=1}^{d}\int_{D_{oy}} v_j(y)\int_{D_{o\tilde{x}}} Q_o(x)\sum_{i=1}^{m} F_{ij}(x)(\partial V(x;v)/\partial x_i)dyd\tilde{x}\end{aligned} \tag{2.82}$$

In order to enable us to make use of Lemma 2.1, we let

$$A = \{V(\cdot;v) : v \in U\} \cap \mathcal{D}_o \tag{2.83}$$

where for each $v \in U$, $V(\cdot;v)$ is the corresponding solution to (2.71)-(2.72). Put $\psi = \mathcal{L}^*Q_o$. Then (2.75), (2.82) and (2.83) yield

$$\begin{aligned}&\sup_{V \in A} \langle \psi, V(\cdot;v) \rangle \\ &= \sup_{v \in U_o} \{-\sum_{j=1}^{d}\int_{D_{oy}} v_j(y)\int_{D_{o\tilde{x}}} Q_o(x)\sum_{i=1}^{m} F_{ij}(x)(\partial V(x;v)/\partial x_i)d\tilde{x}dy\}\end{aligned} \tag{2.84}$$

where

$$U_o \triangleq \{v \in U : V(\cdot;v) \in \mathcal{D}_o \text{ and satisfies (2.71)-(2.72)}\} \subset U. \tag{2.85}$$

In order to satisfy (2.76) it is necessary to take $V_d(x)=1$ for all $x \in D_o$ and to choose Q_o such that

$$\mathcal{L}^*Q_o(x) = 1 - V(x;v^o) \quad \text{a.e. in } D \tag{2.86}$$

$$Q_o(x) = 0 \quad , \quad x \in K \cup D_o^c, \tag{2.87}$$

where $v^o \in U$ is determined by (2.84).

Define

$$\Phi(v) \triangleq \langle \psi, V(\cdot;v) \rangle , \quad v \in U_o \tag{2.88}$$

and

$$\psi_{o,j}(y;v) \triangleq \int_{D_{o\tilde{x}}} Q_o(x) \sum_{i=1}^{m} F_{ij}(x)(\partial V(x;v)/\partial x_i)d\tilde{x}, \; j=1,\ldots,d, y \in D_{oy}, \tag{2.89}$$

then

$$\Phi(v) = \langle \psi, V(\cdot;v) \rangle = - \sum_{j=1}^{d} \int_{D_{oy}} v_j(y)\psi_{o,j}(y;v)dy. \tag{2.90}$$

The following theorem is a straightforward conclusion of this sec= tion.

<u>*Theorem 2.3*</u>

Suppose that $V_o \in \mathcal{D}_o$, $v^o \in U_o$, and Q_o satisfy

$$\mathcal{L}(v^o)V_o(x) = 0, \quad x \in D \tag{2.91}$$

$$V_o(x) = 1, \quad x \in K; \quad V_o(x) = 0, \; x \notin D_o \tag{2.92}$$

$$\mathcal{L}^*Q_o(x) = 1 - V_o(x), \quad \text{a.e. in } D \tag{2.93}$$

$$Q_o(x) = 0, \qquad x \in K \cup D_o^c \tag{2.94}$$

where $v^o \in U_o$ is determined by

$$v^o = \arg\sup_{v \in U_o} \{- \sum_{j=1}^{d} \int_{D_{oy}} v_j(y)\psi_{o,j}(y;v)dy\}; \tag{2.95}$$

then

$$L(v^o) \leq L(v) \text{ for any } v \in U_o. \tag{2.96}$$

Theorem 2.3 states sufficient conditions for the minimization of $L(v)$ on U_o. These conditions can be weakened in the case where c=0 in eqn (2.1).

Let $V : \bar{D}_o \to \mathbb{R}$. We write $V \in W^2(D_o)$ if there exists a sequence of functions $V^{(n)} \in C^2(\bar{D}_o)$ ($C^2(\bar{D}_o)$ denotes the class of all functions $V : D_o \to \mathbb{R}$ such that $\partial^2 V/\partial x_i \partial x_j$, i,j=1,...,m are continuous in D_o and such that V and $\partial V/\partial x_i$ and $\partial^2 V/\partial x_i \partial x_j$, i,j=1,...,m have extensions con= tinuous in $\bar{D}_o$) such that

$$\lim_{n \to \infty} \sup_{x \in \bar{D}_o} |V(x) - V^{(n)}(x)| = 0 \tag{2.97}$$

and

$$\| V^{(k)} - V^{(\ell)} \|_{W^2(D_o)} \to 0 \text{ as } k,\ell \to \infty \quad , \tag{2.98}$$

where, for $f : \bar{D}_o \to \mathbb{R}$

$$\begin{aligned} \|f\|_{W^2(D_o)} \triangleq & \sum_{i,j=1}^{m} \left(\int_{D_o} |\partial^2 f(x)/\partial x_i \partial x_j|^m \, dx\right)^{1/m} \\ & + \sum_{i=1}^{m} \left(\int_{D_o} |\partial f(x)/\partial x_i|^m \, dx\right)^{1/m} + \sup_{x \in \bar{D}_o} |f(x)|. \end{aligned} \tag{2.99}$$

Let $V \in W^2(D_o)$. Then from a Theorem of Krylov ([12], p.122) it follows that, for $v \in U$,

$$V(x) = - E_x^v \int_0^{\tau'} \mathcal{L}(v)V(\zeta_x^v(t))dt + E_x^v V(\zeta_x^v(\tau')) \tag{2.100}$$

where τ' can be taken as $\tau' = \tau_o$ or $\tau' = \tau$, and $\mathcal{L}(v)$ is given by (2.16), where $c(x,u) = 0$ for all $x,u \in \mathbb{R}^m$. Define, for $V \in \mathcal{D}_o$ and $v \in U$:

$$\mathcal{L}(v)V(x) \triangleq \sum_{i=1}^{m} [f_i(x) + \sum_{j=1}^{d} F_{ij}(x)v_j(y)]\partial V(x)/\partial x_i$$
$$+ (\tfrac{1}{2}) \sum_{i,j=1}^{m} (\sigma(x)\sigma'(x))_{ij}\ \partial^2 V(x)/\partial x_i \partial x_j \tag{2.101}$$

and

$$\mathcal{L}^*Q(x) \triangleq - \sum_{i=1}^{m} \partial[f_i(x)Q(x)]/\partial x_i$$
$$+ \tfrac{1}{2} \sum_{i,j=1}^{m} \partial^2[(\sigma(x)\sigma'(x))_{ij}\ Q(x)]/\partial x_i \partial x_j \tag{2.102}$$

for any Q such that $\mathcal{L}^*Q \in L_2(D_0)$.

Denote by U_1 the class of all strategies $v \in U$ such that: there exists a sequence $v^{(n)} \in U_0$, n=1,2,... which converges to v, as $n \to \infty$, in the following sense:

(i) $V(\cdot;v^{(n)})$ converges (via (2.97)-(2.98)) to $V(\cdot;v) \in W^2(D_0)$, as $n \to \infty$;

(ii) $\lim_{n \to \infty} \int_{D_0} |\mathcal{L}(v^{(n)})V(x;v^{(n)})|^m\, dx = \int_{D_0} |\mathcal{L}(v)V(x;v)|^m\, dx = 0$;
where $\mathcal{L}(v)$ is given by (2.101).

(iii) $\sup_{x \in D} E_x^v \tau(x;v) < \infty$ (and consequently $\tau(x;v) < \infty$ P_x^v-almost surely for all $x \in D_0$).

Hence from (i) and (ii) it follows that, for $v \in U_1$

$$\mathcal{L}(v)V(x;v) = 0 \quad \text{a.e. in } D \tag{2.103}$$

$$V(x;v) = 1, \quad x \in K; \qquad V(x;v) = 0, \quad x \notin D_0, \tag{2.104}$$

where $\mathcal{L}(v)$ is given by (2.101). Obviously $U_0 \subset U_1 \subset U$.

The following lemmas will be needed in order to derive sufficient conditions for the minimization of L(v) on U_1.

Lemma 2.2

Consider the stochastic system given by

$$dx = [f(x) + F(x)v(y)]dt + \sigma(x)dW,\ t > 0,\ x \in \mathbb{R}^m \tag{2.105}$$

$$y = (x_1,\dots,x_p),\quad p < m \tag{2.106}$$

where f,F and σ satisfy the conditions stated in Section 2.5, and $v \in U_1$. Let $N \subset D$ be a set of Lebesgue measure 0. Assume:

(a) the transition function $P_v(t,x,\Gamma) = P_x^v(\{\zeta_x^v(t) \in \Gamma\})$, $t \geq 0$, $x \in \mathbb{R}^m$, $\Gamma \in \mathcal{B}(\mathbb{R}^m)$, has a density $P^v(t,x,z)$;

Then

$$P_x^v(\{\Lambda\{t : 0 \leq t \leq \tau(x;v), \zeta_x^v(t) \in N\} = 0\}) = 1 \text{ for all } x \in D, \tag{2.107}$$

where Λ is the Lebesgue measure on the real line.

Proof

Define the following sequence of non-negative random variables:

$$a_n \overset{\Delta}{=} \Lambda\ \{t : 0 \leq t \leq \tau(x;v) \wedge n,\ \zeta_x^v(t) \in N\},\ n=1,2,\dots \tag{2.108}$$

where the notation $a \wedge b = \min(a,b)$ for $a,b \in \mathbb{R}$, is being used. Let

Let $\varepsilon > 0$, then from the Markov inequality we obtain

$$P_x^v(\{a_n > \varepsilon\}) \leq \varepsilon^{-1} E_x^v\, a_n \leq \varepsilon^{-1} E_x^v\, a^* < \infty \tag{2.109}$$

where

$$a^* \overset{\Delta}{=} \Lambda\ \{t : 0 \leq t \leq \tau(x;v),\ \zeta_x^v(t) \in N\}. \tag{2.110}$$

Since a_n can be written in the form

$$a_n = \int_0^{\tau(x;v) \wedge n} I_N(\zeta_x^v(t))dt \tag{2.111}$$

where, for any $A \in \mathcal{B}(\mathbb{R}^m)$

$$I_A(x) \triangleq \begin{cases} 1 & \text{if } x \in A \\ 0 & \text{if } x \notin A \end{cases} \tag{2.112}$$

it follows that

$$\begin{aligned} E_x^v\, a_n &= E_x^v \int_0^{\tau(x;v)\wedge n} I_N(\zeta_x^v(t))dt \le E_x^v \int_0^n I_N(\zeta_x^v(t))dt \\ &= \int_0^n E_x^v\, I_N(\zeta_x^v(t))dt = \int_0^n P_x^v(\{\zeta_x^v(t) \in N\})dt \\ &= \int_0^n \int_N P^v(t,x,z)dz\, dt = 0. \end{aligned} \tag{2.113}$$

Thus (2.109) and (2.113) imply

$$P_x^v(\{a_n > \varepsilon\}) = 0 \text{ for any } \varepsilon > 0 \text{ and } x \in D, \tag{2.114}$$

and consequently

$$\sum_{n=1}^{\infty} P_x^v(\{a_n > \varepsilon\}) < \infty \quad \text{for any } \varepsilon > 0. \tag{2.115}$$

Hence by using a Theorem from Neveu ([92] pp. 46-49) (2.115) implies

$$P_x^v(\{\lim_{n\to\infty} a_n = 0\}) = 1, \tag{2.116}$$

from which (2.107) follows. □

Lemma 2.3

Given $v \in U_1$. Suppose that condition (a) of Lemma 2.2 is satisfied, and let $V(\cdot;v) \in W^2(D_o)$, be a solution to

$$\mathcal{L}(v)V(x;v) = 0, \text{ a.e. in } D \tag{2.117}$$

$$V(x;v) = 1,\ x \in K;\ V(x;v) = 0,\ x \notin D_o, \tag{2.118}$$

where $\mathcal{L}(v)$ is given by (2.101). Then

$$V(x;v) = P_x^v(\{\zeta_x^v(\tau(x;v)) \in K\}),\ x \in D, \tag{2.119}$$

where the family of strong Markov processes $\{(\zeta_x^v, P_x^v),\ x \in \mathbb{R}^m\}$ is deter= mined by (2.105)-(2.106).

Proof

Assume that

$$\mathcal{L}(v)V(x;v) = 0,\ x \in D - N\ ,\ N \subset D \tag{2.120}$$

where N is a set of Lebesgue measure 0. Then from (2.100) and (2.120) it follows that

$$\begin{aligned} V(x;v) &= - E_x^v \int_0^\tau [I_N(\zeta_x^v(t))\mathcal{L}(v)V(\zeta_x^v(t);v) \\ &\quad + (1 - I_N(\zeta_x^v(t)))\mathcal{L}(v)V(\zeta_x^v(t);v)]dt + E_x^v V(\zeta_x^v(\tau);v) \\ &= - E_x^v \int_0^\tau I_N(\zeta_x^v(t))\mathcal{L}(v)V(\zeta_x^v(t);v)dt + E_x^v V(\zeta_x^v(\tau);v) \end{aligned} \tag{2.121}$$

where $\tau = \tau(x;v)$. The proof is completed by making use of (2.107). □

The following theorem is arrived at by applying Lemma 2.3 and fol= lowing the same procedure as in the proof of Theorem 2.3.

Theorem 2.4

Assume that condition (a) of Lemma 2.2 is satisfied for any $v \in U_1$, and suppose that $V_0 \in W^2(D_0)$, $v^0 \in U_1$, and Q_0 satisfy

$$\mathcal{L}(v^0)V_0(x) = 0,\ \text{a.e. in } D \tag{2.122}$$

$$V_0(x) = 1,\quad x \in K;\quad V_0(x) = 0,\quad x \notin D_0 \tag{2.123}$$

$$\mathcal{L}^*Q_0(x) = 1 - V_0(x),\quad \text{a.e. in } D_0 \tag{2.124}$$

$$Q_0(x) = 0,\ x \notin D_0 \tag{2.125}$$

where $v^o \in U_1$ is determined by

$$v^o = \arg\sup_{v \in U_1} \{- \sum_{j=1}^{d} \int_{D_{oy}} v_j(y)\psi_{o,j}(y;v)dy\}; \qquad (2.126)$$

($\mathcal{L}(v)$ and $\mathcal{L}^*Q$ are given by (2.101) and (2.102) respectively, and ψ_o is given by (2.89)). Then

$$L(v^o) \leq L(v) \text{ for any } v \in U_1, \qquad (2.127)$$

where the family $\{(\zeta_x^v, P_x^v),\ x \in \mathbb{R}^m\}$ is determined by (2.105)-(2.106).

Theorem 2.4 states sufficient conditions for the minimization of $L(v)$ on U_1, $U_o \subset U_1 \subset U$. Although, $V(x;v)$ is defined properly by (2.68) for any $v \in U$, we are interested here only in the cases where $V(\cdot;v) \in \mathcal{D}_o$ satisfies eqns (2.71)-(2.72) (the case where $c \neq 0$), and then $v \in U_o$; or in the cases where $V(\cdot;v) \in W^2(D_o)$ satisfies eqns (2.117)-(2.118) (the case where $c=0$), and then $v \in U_1$.

The determination of v^o by means of (2.95) (or by means of (2.126)) is in itself a very difficult optimization problem, and since further= more the establishment of conditions for the existence of solutions $\{V_o, Q_o, v^o\}$ to the complicated eqns (2.91)-(2.95) (or (2.122)-(2.126)) seems to be even more difficult and there is a lack of any background in the theory of partial differential equations, upon which to build, these problems are not considered here.

A procedure for computing weak suboptimal strategies is suggested in the next section.

2.7 COMPUTATION OF WEAK SUBOPTIMAL STRATEGIES

In this work the following algorithm has been applied to eqns (2.91)-(2.95) (or (2.122)-(2.126)) in order to compute weak suboptimal strategies.

1. Given $v^{(o)}, v^{(1)}, \dots, v^{(n)} \in U_o$ (or U_1 in the case of (2.122)-(2.126)).

2. Compute $V(\cdot;v^{(n)})$ by solving numerically the following problem:

$$\mathcal{L}(v^{(n)})V(x) = 0 \quad , \quad x \in D \qquad \text{(or a.e. in D)} \tag{2.128}$$

$$V(x) = 1 \quad , \quad x \in K; \quad V(x) = 0 \quad , \quad x \notin D_o \tag{2.129}$$

3. Calculate $L(v^{(n)})$.

4. Compute $Q(\cdot;v^{(n)})$ by solving numerically the following problem:

$$\mathcal{L}^*Q(x) = 1 - V(x;v^{(n)}), \quad \text{a.e. in D (or a.e. in } D_o) \tag{2.130}$$

$$Q(x) = 0 \quad , \quad x \in K \cup D_o^c \quad (\text{or } x \notin D_o). \tag{2.131}$$

5. Compute $v^{(n+1)}$ by

$$v_j^{(n+1)}(y) = -v_{o,j}\text{sign}\{\int_{D_{o\tilde{x}}} Q(x;v^{(n)}) \sum_{i=1}^{m} F_{ij}(x)(\partial V(x;v^{(n)})/\partial x_i)d\tilde{x}\}, j=1,\dots,d, y \in D_{oy}. \tag{2.132}$$

6. If $v^{(n+1)} \neq v^{(n)}$; then $v^{(n+1)} \to v^{(n)}$, and go to 2. Otherwise: stop.

The computations are continued until for some $n \geq 0$ either $v^{(n+1)} = v^{(n)}$ or $L(v^{(n+1)}) = L(v^{(n)})$.

Remark 2.4

If the sequence $\{v^{(n)}\}$ converges, and $\lim_{n \to \infty} v^{(n)} = \bar{v}$, then $\{V(\cdot;\bar{v}), Q(\cdot;\bar{v}), \bar{v}\}$ is a solution to the following problem:

$$\mathcal{L}(v)V(x) = 0 \quad , \quad x \in D \qquad \text{(or a.e. in D)} \tag{2.133}$$

$$\mathcal{L}^*Q(x) = 1 - V(x) \quad , \quad \text{a.e. in D} \quad (\text{or a.e. in } D_o) \tag{2.134}$$

$$V(x) = 1,\ x \in K; \quad V(x) = Q(x) = 0,\ x \notin D_o; \quad Q(x) = 0,\ x \in K \quad \text{(not valid in the case c=0)} \tag{2.135}$$

$$v_j(y) = -v_{o,j}\operatorname{sign}\{\int_{D_{o\tilde{x}}} Q(x) \sum_{i=1}^{m} F_{ij}(x)(\partial V(x)/\partial x_i)d\tilde{x}\},\ j=1,\dots,d,\ y \in D_{oy}. \tag{2.136}$$

For all the examples which have been numerically solved here, the results indicate that $L(v^{(o)}) \geq L(v^{(1)}) \geq \dots \geq L(v^{(n)}) \geq \dots$.

Unfortunately, owing to the complexity of eqns (2.91)-(2.95) ((2.122)-(2.126)), and of the algorithm for computing weak suboptimal strategies, we cannot here give conditions for the existence of $\lim_{n \to \infty} v^{(n)} = \bar{v}$, and whenever $\{v^{(n)}\}$ converges to $\bar{v}$ we cannot determine whether $\bar{v}$ is also a weak optimal strategy. Nevertheless, the numerical results presented in this chapter, and other chapters as well, for the examples which have been numerically solved, suggest that $\bar{v}$, whenever it exists, is a good approximation to an optimal strategy.

2.8 THE NUMERICAL METHOD

In this section, a finite-difference scheme is described for solving numerically (2.128)-(2.129) (for a given strategy $v \in U_o$) and (2.130)-(2.131), where $\mathcal{L}(v)$ and $\mathcal{L}^*$ are given by (2.16) and (2.79) respectively. It is assumed that f, F, σ, c, P_J and M satisfy all the conditions stated in Section 2.5. It is further assumed that $(\sigma(x)\sigma'(x))_{ij} = \delta_{ij}\ \sigma_{ii}^2(x),\ \ x \in \mathbb{R}^m,\ \ i,j=1,\dots,m,$ which is the case for all the examples here solved numerically.

Let $\mathbb{R}_h^m$ be a finite-difference grid on $\mathbb{R}^m$, with a constant mesh size h along all axes. Define $D_h \triangleq \mathbb{R}_h^m \cap D$, $D_{oh} \triangleq \mathbb{R}_h^m \cap D_o$ and $K_h \triangleq \mathbb{R}_h^m \cap K$. Denote by e^i the unit vector along the i-th axis, i=1,...,m. Then (2.128) and (2.130) are replaced by

$$\sum_{i=1}^{m} F_i(x;v)d_iV(x) + \sum_{i=1}^{m} \sigma_{ii}^2(x)(V(x+e^ih) + V(x-e^ih) - 2V(x))/(2h^2)$$
$$+ \rho(PV)(x) - \rho V(x) = 0,\ \ v \in U_o,\ \ x \in D_h \tag{2.137}$$

and

$$\sum_{i=1}^{m}(-f_i(x) + d_i^c\,\sigma_{ii}^2(x))d_iQ(x) + \sum_{i=1}^{m}(-d_i^c\,f_i(x) + (\tfrac{1}{2})(d_i^c)^2\,\sigma_{ii}^2(x)$$

$$-(\rho/m))Q(x) + \sum_{i=1}^{m}\sigma_{ii}^2(x)(Q(x+e^ih) + Q(x-e^ih) - 2Q(x))/(2h^2) \quad (2.138)$$

$$+\rho(P^*Q)(x) = 1 - V(x) \quad , \quad x \in D_h$$

respectively, where

$$F_i(x;v) \triangleq f_i(x) + \sum_{j=1}^{d} F_{ij}(x)v_j(y), \; i=1,\ldots,m, \quad x \in D_h \qquad (2.139)$$

$$d_i^c\,g(x) \triangleq (g(x+e^ih) - g(x-e^ih))/(2h), \; i=1,\ldots,m \qquad (2.140)$$

$$\psi(x)d_iV(x) \triangleq \begin{cases} \psi(x)(V(x+e^ih) - V(x))/h & \text{if } \psi(x) \geq 0 \\ \\ \psi(x)(V(x) - V(x-e^ih))/h & \text{if } \psi(x) < 0 \end{cases} \qquad (2.141)$$

$$(PV)(x) \triangleq \int_{\mathbb{R}^m} V(x + c(x,u))P_J(du), \quad x \in D_h, \qquad (2.142)$$

$$(P^*Q)(x) \triangleq \int_{\mathbb{R}^m} Q(C(x,u))\Delta(x,u)P_J(du), \quad x \in D_h \text{ (see (2.21))}. \qquad (2.143)$$

$\{d_iQ\}$ are defined in the same manner as in (2.141). The function ψ stand for $F_i(x;v)$ or $(-f_i(x) + d_i^c\,\sigma_{ii}^2(x))$, $i=1,\ldots,m$, as the case may be. Elimination now yields

$$V(x) = (FV)(x), \quad x \in D_h \qquad (2.144)$$

$$V(x) = 1, \quad x \in K_h; \quad V(x) = 0, \quad x \notin D_{oh} \qquad (2.145)$$

and

$$Q(x) = (G\{Q,V\})(x), \quad x \in D_h \qquad (2.146)$$

$$Q(x) = 0, \quad x \in K_h \cup D_{oh}^c \qquad (2.147)$$

where

$$(FV)(x) \triangleq \sum_{i=1}^{m} (P_i(x;v)V(x+e^i h) + P_{-i}(x;v)V(x-e^i h))$$

$$+ h^2\rho(PV)(x)/R(x;v) \quad , \quad x \in D_h \tag{2.148}$$

$$R(x;v) \triangleq \sum_{i=1}^{m} \sigma_{ii}^2(x) + \rho h^2 + h \sum_{i=1}^{m} |F_i(x;v)|, \quad x \in D_h \tag{2.149}$$

$$P_i(x;v) \triangleq [\sigma_{ii}^2(x)/2 + h \max(F_i(x;v),0)]/R(x;v)$$

$$i=1,\dots,m \quad , \quad x \in D_h \tag{2.150}$$

$$P_{-i}(x;v) \triangleq [\sigma_{ii}^2(x)/2 - h \min(F_i(x;v),0)]/R(x;v)$$

$$i=1,\dots,m \quad , \quad x \in D_h \tag{2.151}$$

and

$$(G\{Q,V\})(x) \triangleq \sum_{i=1}^{m} (S_i(x)Q(x+e^i h) + S_{-i}(x)Q(x-e^i h))$$

$$+ h^2\rho(P^*Q)(x)/S(x) - h^2(1 - V(x))/S(x), \quad x \in D_h \tag{2.152}$$

$$S(x) \triangleq \sum_{i=1}^{m} (\sigma_{ii}^2(x) - h^2 b_i(x) + h|a_i(x)|), \quad x \in D_h \tag{2.153}$$

$$a_i(x) \triangleq -f_i(x) + d_i^c\, \sigma_{ii}^2(x), \ i=1,\dots,m, \quad x \in D_h \tag{2.154}$$

$$b_i(x) \triangleq -d_i^c\, f_i(x) + (\tfrac{1}{2})(d_i^c)^2\, \sigma_{ii}^2(x) - (\rho/m), \ x \in D_h \tag{2.155}$$

$$S_i(x) \triangleq [\sigma_{ii}^2(x)/2 + h \max(a_i(x),0)]/S(x)$$

$$i=1,\dots,m \quad , \quad x \in D_h \tag{2.156}$$

$$S_{-i}(x) \triangleq [\sigma_{ii}^2(x)/2 - h \min(a_i(x),0)]/S(x)$$

$$i=1,\dots,m, \quad x \in D_h. \tag{2.157}$$

Eqns (2.144)-(2.145), and eqns (2.146)-(2.147) (for a given V) are solved by an iterative procedure using the underrelaxation technique with an acceleration factor w_o, until the difference between two consecutive iterations does not exceed a given tolerance ε_o.

Given $h > 0$, we assume that the algorithm for computing weak suboptimal strategies, where the scheme given by (2.144)- (2.157) is applied, converges to a unique solution denoted here by $(V_o^h, Q_o^h, \bar{v}^h)$. Then the value of $C^h(\bar{v}^h)$, the approximation to $C(v^o)$ (see (2.65) for the definition of $C(v)$), can be computed by

$$C^h(\bar{v}^h) = \int_{D_o} V_o^h(x)\mu(dx). \tag{2.158}$$

Remark 2.5

Let $v \in U_o$ be given. Then by using probabilistic techniques, conditions can be established on $V^h(\cdot;v)$, the solution to (2.144)-(2.145), for the convergence of $V^h(\cdot;v)$ to $V(\cdot;v)$, as $h \downarrow 0$. For more details see Kushner and Dimasi [93], Kushner [94] and the references cited there.

2.9 EXAMPLE 1 : A PATROL PROBLEM

The following example is taken from Yavin and Reuter [95].

2.9.1 Statement of the problem

Consider the random motion of two points M_e and M_p in an open and bounded domain A_o in the plane, each of the velocities (v_1,v_2) of M_e and $(v_3,0)$ of M_p, being perturbed by an $\mathbb{R}^2$-valued and an $\mathbb{R}$-valued Gaussian white noise respectively. Assume given a closed target set $K_o \subset A_o$. At any instant, the velocity (v_1,v_2) of M_e is directed towards the centre of K_o. The point M_p moves along a section $(-\ell,\ell) \times \{d\}$ in A_o.

It is assumed that M_p can observe only its own location and cannot observe the location of M_e. The goal of M_p is to intercept M_e, in A_o, before M_e reaches the target set K_o.

More precisely, the motion of M_e is described by

$$dx_i = v_i(\tilde{x})dt + \sigma_i dW_i, \quad t > 0, \quad i=1,2, \tag{2.159}$$

and the motion of M_p is given by

$$dx_3 = v_3(y)dt + \sigma_3 dW_3 \ , \quad t > 0 \tag{2.160}$$

$$dx_4 = 0 \quad , \quad t > 0, \quad x_4(0) = d \tag{2.161}$$

where $\tilde{x} = (x_1,x_2)$ and $y = (x_3,x_4)$ are the coordinates of M_e and M_p respectively, and

$$v_i(\tilde{x}) = \begin{cases} -u_o\, x_i\, [x_1^2 + x_2^2]^{-\frac{1}{2}} & \tilde{x} \in \mathbb{R}^2 - K_o \\ & \qquad\qquad i=1,2 \\ f_i(\tilde{x}) & \tilde{x} \in K_o \end{cases} \tag{2.162}$$

$$\tilde{K}_o \triangleq \{\tilde{x} : x_1^2 + x_2^2 \le \rho^2\} \times [\, -\ell + \varepsilon, \ell - \varepsilon] = K_o \times [-\ell + \varepsilon, \ell - \varepsilon] \tag{2.163}$$

and σ_i, i=1,2,3; u_o, ρ, ℓ and ε are given positive numbers; $\rho + \varepsilon < |d| < \ell - \varepsilon$. The functions f_i, i=1,2 are taken to be such that v_i, i=1,2, are bounded and continuously differentiable on $\mathbb{R}^2$. $W = \{W(t) = (W_1(t),W_2(t),W_3(t)), t \ge 0\}$ is an $\mathbb{R}^3$-valued standard Wiener process. The strategy $v_3 = v_3(x_3)$ $(x_4(t) = d, t \ge 0$, and d is considered here as a given parameter) is assumed here to be an element of U_1. Denote by $\zeta_x^v = \{\zeta_x^v(t) = (\zeta_{x,1}^v(t), \zeta_{x,2}^v(t), \zeta_{x,3}^v(t)), t \ge 0\}$, $x = (x_1,x_2,x_3)$, $v = (v_1,v_2,v_3)$ the (weak) solution to (2.159)-(2.160). Let

$$D_o \triangleq \{x : |x_i| < \ell \ , \ i = 1,2,3\} \tag{2.164}$$

$$D_1 \triangleq \{x : |x_i| \leq \ell - \varepsilon, \; i=1,2,3 \text{ and } x_1^2 + x_2^2 \geq (\rho+\varepsilon)^2\}. \tag{2.165}$$

If a certain $t \geq 0$ is the first time that

$$(\zeta^v_{x,1}(t) - \zeta^v_{x,3}(t))^2 + (\zeta^v_{x,2}(t) - d)^2 \leq \varepsilon^2 \text{ and } \zeta^v_x(t) \in D_1, \tag{2.166}$$

then it is said that M_e has been intercepted by M_p in the set D_1.

Define the sets K and D as

$$K \triangleq \{x : (x_1 - x_3)^2 + (x_2 - d)^2 \leq \varepsilon^2 \text{ and } x \in D_1\} \tag{2.167}$$

$$D \triangleq D_0 - (\tilde{K}_0 \cup K) \tag{2.168}$$

and let $\tau(x;v)$ be the first exit time of ζ^v_x from D as defined in (2.6).

The problem considered in this example is to maximize the functional $P^v(\&)$, (2.66), on U_1 where the event & is defined as

$$\& \triangleq \{M_p \text{ intercepts } M_e \text{ in } D_1, \text{ before } M_e \text{ reaches the set } K_0 \text{ and before } (M_e,M_p) \text{ has left } D_0\}. \tag{2.169}$$

Then

$$P^v(\&) = \int_{D_0} P^v_x(\&)\mu(dx) = \int_{D_0} P^v_x(\{\zeta^v_x(\tau(x;v)) \in K\})\mu(dx) = C(v) \tag{2.170}$$

where

$$\mu(B) = P^v_x(\{\zeta^v_x(0) \in B\}), \; B \in \mathcal{B}(\mathbb{R}^3).$$

Assuming that condition (a) of Lemma 2.2 is satisfied for any $v \in U_1$ we apply the procedure for computing weak suboptimal strategies. In this case, eqns (2.128)-(2.129), and (2.130)-(2.131) take the following form:

$$\sum_{i=1}^{2} v_i(\tilde{x})\partial V(x)/\partial x_i + v_3^{(n)}(y)\partial V(x)/\partial x_3 + (\tfrac{1}{2})\sum_{i=1}^{3}\sigma_i^2\,\partial^2 V(x)/\partial x_i^2 = 0,\ x \in D, \tag{2.171}$$

$$-\sum_{i=1}^{2}\partial[v_i(\tilde{x})Q(x)]/\partial x_i + (\tfrac{1}{2})\sum_{i=1}^{3}\sigma_i^2\,\partial^2 Q(x)/\partial x_i^2 = 1 - V(x;v^{(n)}) \text{ a.e. in } D_o \tag{2.172}$$

$$V(x) = 1,\ x \in K;\ V(x) = 0,\ x \in \tilde{K}_o \cup D_o^c;\ Q(x) = 0,\ x \in D_o^c, \tag{2.173}$$

while (2.132) yields

$$v_3^{(n+1)}(x_3) = -v_o \operatorname{sign}\left(\int_{-\ell}^{\ell}\int_{-\ell}^{\ell} Q(x;v^{(n)})(\partial V(x;v^{(n)})/\partial x_3)dx_1\, dx_2\right) \tag{2.174}$$

$x_3 \in (-\ell,\ell)$

where $v^{(n)} = (v_1,v_2,v_3^{(n)})$, n=0,1,2,... . Note that $|v_3(x_3)| \le v_{o,3} = v_o$ for all $x_3 \in (-\ell,\ell)$.

2.9.2 Results

Throughout this subsection, the following forms for $\mu(dx)$ were taken:

$$\mu^{(1)}(dx) = \begin{cases} dx_1 dx_2 dx_3/[(2\ell L)(2\ell)] & (x,x_4) \in C_1 \\ 0 & \text{otherwise} \end{cases} \tag{2.175}$$

and

$$\mu^{(2)}(dx) = \begin{cases} dx_1 dx_2 dx_3/[(2L^2)(2\ell)] & (x\ ,x_4) \in C_2 \\ 0 & \text{otherwise} \end{cases} \tag{2.176}$$

where

$$C_1 = (-\ell,\ell) \times [\ell - L,\ell) \times (-\ell,\ell) \times \{d\},\quad 0 < L < \ell \tag{2.177}$$

$$C_2 = [-L,L] \times [\ell-L,\ell) \times (-\ell,\ell) \times \{d\}. \tag{2.178}$$

The distribution $\mu^{(1)}$ represents the case where it is known that at $t=0$, the position of M_e has a uniform distribution on the strip $(-\ell,\ell) \times [\ell-L,\ell)$ located at the boundary of D_0, and that the position of M_p has a uniform distribution along the section $(-\ell,\ell) \times \{d\}$. $\mu^{(2)}$ represents the case where more information is available on the position of M_e at $t=0$. In this case the position of M_e has a uniform distribution on a 'window' $[-L,L] \times [\ell-L,\ell)$ located at the boundary of D_0, and the position of M_p has a uniform distribution along the section $(-\ell,\ell) \times \{d\}$.

Computations were carried out for $\sigma_1^2 = \sigma_2^2 = \sigma_3^2 = 0.05$, $\ell=1$, $\rho=0.1$, $L=0.2$, $\varepsilon=10^{-4}$, $u_0=0.1$, $v_0/u_0=0.5,1,2,4,8$, $\rho+\varepsilon < d \le 0.8$, $\varepsilon_0=10^{-3}$, and $h=0.1$. Figs. 2.1 and 2.2 show $C^h(\bar{v}^h)$ (2.158) as a function of d and for various values of v_0/u_0. The distribution $\mu^{(1)}$ is used in Fig.2.1, the distribution $\mu^{(2)}$ in Fig.2.2. The results show that $C^h(\bar{v}^h)$ decreases when d increases, and that $C^h(\bar{v}^h)$ increases when v_0/u_0 increases. In all the cases computed here the procedure for computing weak suboptimal strategies converged and it was found that $\bar{v}^h$, the limit to $\{v^{(n)}\}$, is given by $\bar{v}^h(x_3) = -v_0 \operatorname{sign}(x_3)$, $|x_3| < \ell$, where $\operatorname{sign}(0) = 0$. Note that the values of $C^h(\bar{v}^h)$ in Fig. 2.2 are greater than the corresponding values of $C^h(\bar{v}^h)$ in Fig.2.1. This is because $\mu^{(2)}$ represents the case where more information is available on the location of $(\zeta^v_{x,1}(0), \zeta^v_{x,2}(0))$ than when $\mu^{(1)}$ is used.

In order to assess the accuracy of the numerical method used here, three mesh sizes were considered, viz. $h=0.2$, 0.1, 0.05. Tables 2.1-2.3 show the values of V_0^h and Q_0^h at the points $x^{(i)}$, $i=1,\dots,6$, where $x^{(1)} = (\pm 0.2, 0.4, \pm 0.8)$, $x^{(2)} = (0.0, 0.8, \pm 0.6)$, $x^{(3)} = (\pm 0.2, 0.0, \pm 0.6)$, $x^{(4)} = (\pm 0.6, 0.0, \pm 0.4)$, $x^{(5)} = (0.0, 0.0, \pm 0.2)$, and $x^{(6)} = (\pm 0.6, 0.8, 0.0)$. The results given in these tables, and other results as well, suggest that as $h \downarrow 0$, V_0^h and Q_0^h converge to the solutions

$V(\cdot;\bar{v})$ and $Q(\cdot;\bar{v})$ respectively, of (2.133)-(2.136). During the computation, use was made of the symmetry properties of the solutions: $(V(-x_1,x_2,-x_3) = V(x_1,x_2,x_3)$ and $Q(-x_1,x_2,-x_3) = Q(x_1,x_2,x_3))$.

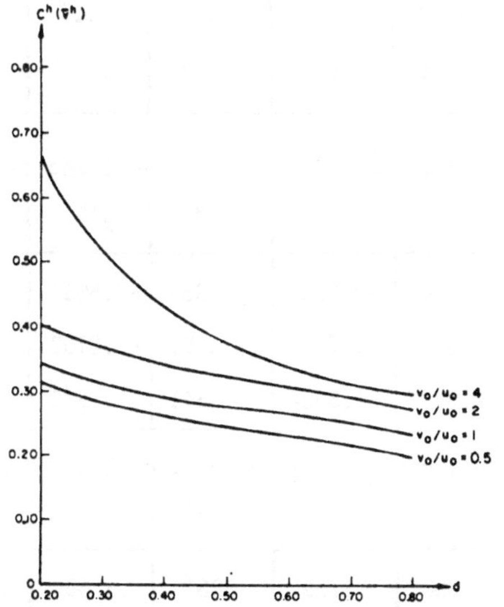

Fig. 2.1: $c^h(\bar{v}^h)$ as a function of d for various values of v_0/u_0, and where $\mu = \mu^{(1)}$.

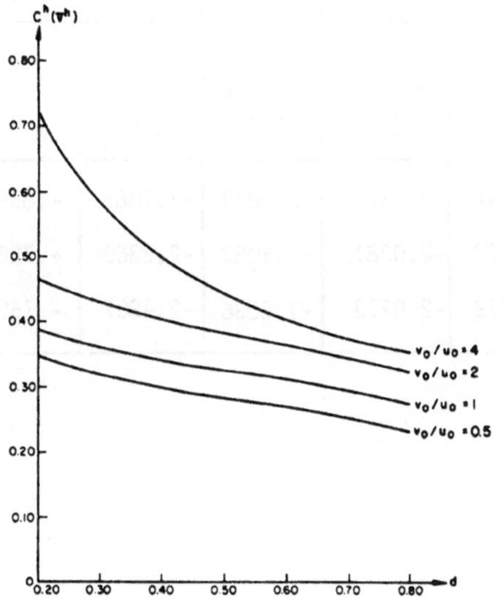

Fig.2.2: $c^h(\bar{v}^h)$ as a function of d for various values of v_0/u_0, and where $\mu = \mu^{(2)}$.

Table 2.1a: The values of $V_o^h(x^{(i)})$ for $u_o=v_o=0.1$ and $d=0.2$

h	$x^{(1)}$	$x^{(2)}$	$x^{(3)}$	$x^{(4)}$	$x^{(5)}$	$x^{(6)}$
0.2	.3188	.2463	.4874	.5306	.6016	.2527
0.1	.4536	.3557	.6962	.7640	.8615	.3490
0.05	.4697	.3756	.7081	.7698	.8733	.3638

Table 2.1b: The values of $Q_o^h(x^{(i)})$ for $u_o=v_o=0.1$ and $d=0.2$

h	$x^{(1)}$	$x^{(2)}$	$x^{(3)}$	$x^{(4)}$	$x^{(5)}$	$x^{(6)}$
0.2	-1.3179	-.9467	-2.2283	-1.6054	-2.5017	-.8413
0.1	-1.0527	-.8300	-1.6127	-1.1501	-1.8102	-.7662
0.05	-1.0424	-.8347	-1.5824	-1.1541	-1.8310	-.7858

Table 2.2a: The values of $V_o^h(x^{(i)})$ for $u_o=v_o=0.1$ and $d=0.4$

h	$x^{(1)}$	$x^{(2)}$	$x^{(3)}$	$x^{(4)}$	$x^{(5)}$	$x^{(6)}$
0.2	.2798	.2380	.3156	.3281	.3759	.2622
0.1	.4265	.3698	.5300	.5462	.6382	.3855
0.05	.4262	.3793	.5180	.5370	.6269	.3933

Table 2.2b: The values of $Q_o^h(x^{(i)})$ for $u_o=v_o=0.1$ and $d=0.4$

h	$x^{(1)}$	$x^{(2)}$	$x^{(3)}$	$x^{(4)}$	$x^{(5)}$	$x^{(6)}$
0.2	-1.4032	-.9740	-2.7175	-2.0039	-3.1088	-.8397
0.1	-1.0792	-.8179	-2.0381	-1.5083	-2.3369	-.7223
0.05	-1.0947	-.8379	-2.0778	-1.5536	-2.4807	-.7523

Table 2.3a: The values of $V_0^h(x^{(i)})$ for $u_0=0.1, v_0=0.3$ and $d=0.4$

h	$x^{(1)}$	$x^{(2)}$	$x^{(3)}$	$x^{(4)}$	$x^{(5)}$	$x^{(6)}$
0.2	.4333	.3251	.3359	.3236	.3143	.2559
0.1	.6387	.4789	.5825	.5371	.5369	.3694
0.05	.6814	.5018	.6267	.5529	.5889	.3759

Table 2.3b: The values of $Q_0^h(x^{(i)})$ for $u_0=0.1, v_0=0.3$ and $d=0.4$

h	$x^{(1)}$	$x^{(2)}$	$x^{(3)}$	$x^{(4)}$	$x^{(5)}$	$x^{(6)}$
0.2	-1.0398	-.8136	-2.3874	-1.7984	-2.7390	-.7669
0.1	- .6323	-.6254	-1.5622	-1.2399	-1.8350	-.6486
0.05	- .6029	-.6309	-1.4823	-1.2371	-1.8282	-.6856

2.10 EXAMPLE 2 : A RENDEZVOUS PROBLEM

The following example is taken from Yavin [96].

2.10.1 Introduction

Consider a random motion of two points M_1 and M_2 in the plane, and suppose that each of the velocities $u = (u_1, u_2)$ of M_1 and $v = (v_1, v_2)$ of M_2, is perturbed by a corresponding $\mathbb{R}^2$-valued Gaussian white noise. Each of the points M_1 and M_2 wishes to rendezvous with the other in the unit square of $\mathbb{R}^2$. If, given $\varepsilon > 0$, a certain $t \geq 0$ is the first time that $d(M_1, M_2) \leq \varepsilon$ (where $d(M_1, M_2)$ denotes the distance between M_1 and M_2) then it is said that M_1 has rendezvoused with M_2.

The points cannot observe each other (unless $d(M_1, M_2) \leq \varepsilon$). Also, each of the points uses a different navigation system, from which it follows that M_1 measures its position $x = (x_1, x_2)$ relative to a fixed coordinate system S_1 in the plane, with origin 0_1; and M_2 measures its position $y = (y_1, y_2)$ relative to another fixed coordinate system S_2 in the plane, with origin 0_2. It is assumed that S_1 is parallel to S_2 (see Fig. 2.3). Denote by 0_0 the origin of $\mathbb{R}^2$. Then $d(0_0, 0_1)$ and $d(0_0, 0_2)$

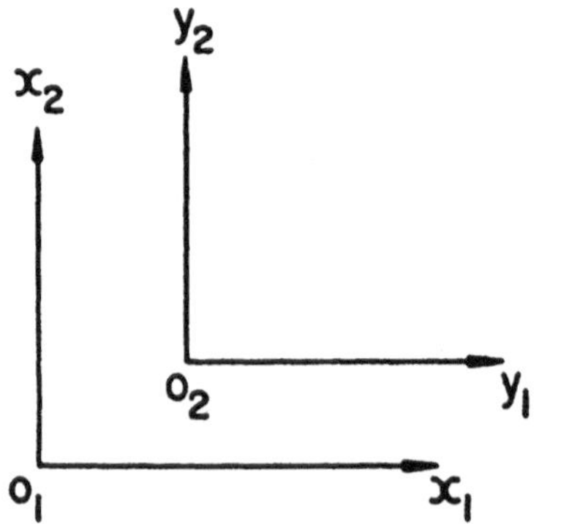

Fig.2.3: The coordinate systems $S_1 = (0_1, x_1, x_2)$ and $S_2 = (0_2, y_1, y_2)$.

represent the navigational errors of M_1 and M_2 respectively. The points M_1 and M_2 are moving in hostile territory, i.e. there exists a killing density $a(\cdot,\cdot)$ such that

$$\begin{aligned}&\text{Prob.}\{\text{Either } M_1 \text{ or } M_2 \text{ is killed in } [t,t+\Delta)|M_1 \text{ and } M_2 \text{ are at}\\ &\text{positions } x \text{ and } y \text{ respectively at time } t\} \qquad (2.179)\\ &= a(x,y)\Delta+o(\Delta), \quad t \geq 0,\ \Delta \geq 0, \quad x,y \in \mathbb{R}^2.\end{aligned}$$

It is assumed here that $a:\mathbb{R}^4 \to \mathbb{R}$ is a given bounded, continuous and non-negative function.

Define the following sets:

$$D_x \triangleq \{x : |x_i| < 1, \quad i=1,2\} ; \tag{2.180}$$

$$D_y \triangleq \{y : |y_i| < 1, \quad i=1,2\} ; \tag{2.181}$$

$$A_x \triangleq \{x : |x_i| \leq 1-\delta, \ i=1,2\} ; \tag{2.182}$$

$$A_y \triangleq \{y : |y_i| \leq 1-\delta, \ i=1,2\} ; \tag{2.183}$$

where $0 < \delta \ll 1$ is given. Assuming that at t=0, $(M_1,M_2) \in D_x \times D_y$, let $\mathcal{E}$ denote the following event: $\mathcal{E} = \{M_1$ rendezvouses with M_2 in $A_x \times A_y$ before either of them is killed and before (M_1,M_2) leaves $D_x \times D_y\}$. The problem dealt with here is to find a pair $(u^*,v^*) = \{(u^*(x),v^*(y)), (x,y) \in \mathbb{R}^4\}$ such that Prob.$(\mathcal{E})$ will be maximized on a class of $(u,v) = \{(u(x),v(y)), (x,y) \in \mathbb{R}^4\}$-admissible velocity laws (strategies).

Denote by 0_o the origin of $\mathbb{R}^2$ and by $c = (c_1,c_2)$ the position of 0_2 relative to S_1. It is assumed here that $0_o = 0_1$. Two cases are treated here: a special case where c is deterministic and known; and the more general case where c is an $\mathbb{R}^2$-valued random element with a given pro= bability measure μ on $\mathcal{B}(\mathbb{R}^2)$(where $\mathcal{B}(\mathbb{R}^2)$ denotes the σ-algebra of Borel sets of $\mathbb{R}^2$). In the first case, where c is deterministic and known, it is possible to fix c=0 without loss of generality.

Owing to the form of the velocity law $(u,v) = \{(u(x),v(y)),(x,y) \in \mathbb{R}^4\}$ it follows that the problem considered here is a problem of the control of a partially observable stochastic system.

2.10.2 Formulation of the Problem

Consider the random motion of the following points, M_1 and M_2, in the plane. The motion of M_1 is given by

$$dx_i = u_i(x)dt + \sigma_i dW_i \quad , \quad i=1,2, \quad t > 0 \tag{2.184}$$

and the motion of M_2 is given by

$$dy_i = v_i(y)dt + \gamma_i dB_i \quad , \quad i=1,2, \quad t > 0 \tag{2.185}$$

where $x = (x_1,x_2)$ are the coordinates of M_1 relative to the coordinate system S_1, and $y = (y_1,y_2)$ are the coordinates of M_2 relative to the coordinate system S_2; σ_i, γ_i, i=1,2 are given positive numbers; and $\{W(t) = (W_1(t), W_2(t)), t \geq 0\}$ and $\{B(t) = (B_1(t), B_2(t)), t \geq 0\}$ are mutually independent $\mathbb{R}^2$-valued standard Wiener processes.

Let $\tilde{U}$ be the class of all measurable functions $\theta = (u,v) = \{(u(x),v(y)), \alpha = (x,y) \in \mathbb{R}^4\}$ such that u and v are measurable and $|u_i(x)| \leq u_0$, i=1,2, $x \in \mathbb{R}^2$ and $|v_i(y)| \leq v_0$, i=1,2, $y \in \mathbb{R}^2$, where u_0 and v_0 are given posi= tive numbers. Given a velocity law (strategy) $\theta \in \tilde{U}$ and an initial condition $\alpha \in \mathbb{R}^4$, it follows from the theory of Stroock and Varadhan [89], that equations (2.184)-(2.185) have a unique weak solution $(\hat{\zeta}_\alpha^\theta(t),\hat{P}_\alpha^\theta)$ which is a strong Markov process.

Although the process $(\hat{\zeta}_\alpha^\theta(t),\hat{P}_\alpha^\theta)$ describes the motion of (M_1,M_2), its behaviour still does not represent the situation where the points are moving in hostile territory. Following Dynkin [81] (Chapters IX and X) let

$$P(\alpha,t,\Gamma;\theta) = \int_{\{\hat{\zeta}_\alpha^\theta(t)\in\Gamma\}} \exp(-\int_0^t a(\hat{\zeta}_\alpha^\theta(s))ds)\hat{P}_\alpha^\theta(d\omega), \ t \geq 0, \quad \Gamma \in \mathcal{B}(\mathbb{R}^4) \tag{2.186}$$

and define

$$\zeta_\alpha^\theta(t) = \hat{\zeta}_\alpha^\theta(t) \qquad 0 \leq t < \tau_k(\alpha;\theta). \tag{2.187}$$

Then (Dynkin [81] Chapters IX and X) P is a transition function on $(\mathbb{R}^4,\mathcal{B}(\mathbb{R}^4))$, and $(\zeta_\alpha^\theta(t),P_\alpha^\theta)$ is a strong Markov process with a transition function given by (2.186). Furthermore, ζ_α^θ is killed (terminates) at the random time $\tau_k(\alpha;\theta)$, and the process

$$\alpha_t = \exp(-\int_0^t a(\hat{\zeta}_\alpha^\theta(s))ds) \tag{2.188}$$

has the following interpretation:

$$P_\alpha^\theta(\{\tau_k(\alpha;\theta) > t\} \mid \hat{\zeta}_\alpha^\theta(s),\ 0 \leq s \leq t) = \exp(-\int_0^t a(\hat{\zeta}_\alpha^\theta(s))ds) . \tag{2.189}$$

where for $B \in \mathcal{B}(\mathbb{R}^4)$,

$$P_\alpha^\theta(\{\zeta_\alpha^\theta(t) \in B\}\cap\{\tau_k(\alpha;\theta) > t\}) = \int_{\{\hat{\zeta}_\alpha^\theta(t) \in B\}} \exp(-\int_0^t a(\hat{\zeta}_\alpha^\theta(s))ds)\hat{P}_\alpha^\theta(d\omega). \tag{2.190}$$

(see also Prohorov and Rozanov [97]).

The process $(\zeta_\alpha^\theta(t),P_\alpha^\theta)$ describes the motion of the points M_1 and M_2 whose kinematics are governed by eqns (2.184)-(2.185) and which are killed at a random time τ_k according to a probability law such as that described by (2.179).

Let $\zeta_\alpha^\theta(t) = (\chi_\alpha^\theta(t),\eta_\alpha^\theta(t)) = (\chi_{x,1}^u(t),\chi_{x,2}^u(t),\eta_{y,1}^v(t),\eta_{y,2}^v(t))$, then at the time t, $d(M_1,M_2) = [\sum_{i=1}^{2} (\chi_{x,i}^u(t) - \eta_{y,i}^v(t) - c_i)^2]^{\frac{1}{2}}$.

Define the following sets in $\mathbb{R}^4$:

$$D_0 \triangleq D_x \times D_y ; \tag{2.191}$$

$$K(c) \triangleq \{\alpha = (x,y) : (x_1-y_1-c_1)^2+(x_2-y_2-c_2)^2 \leq \varepsilon^2 \text{ and } (x,y) \in A_x \times A_y\}, \tag{2.192}$$

$c \in C$;

$$D(c) \triangleq D_o - K(c), \quad c \in C; \tag{2.193}$$

and the family of random times:

$$\tau(\alpha,c;\theta) \triangleq \begin{cases} \inf\{0 \leq t < \tau_k(\alpha;\theta) : \zeta_\alpha^\theta(t) \notin D(c) \text{ when } \zeta_\alpha^\theta(0) = \alpha \in D(c)\} \\ 0 \quad \text{if} \quad \zeta_\alpha^\theta(0) = \alpha \notin D(c) \\ \tau_k(\alpha;\theta) \text{ if } \zeta_\alpha^\theta(t) \in D(c) \text{ for all } t \in [0,\tau_k(\alpha;\theta)] \end{cases} \tag{2.194}$$

$c \in C$, where C is a given centrally symmetric closed and bounded domain in $\mathbb{R}^2$ with the point 0_1 at its centre. It is assumed here that the pro= bability measure μ satisfies : $\mu(B) = 0$ if $B \cap C = \phi$, $B \in \mathcal{B}(\mathbb{R}^2)$, where ϕ denotes the empty set. Let $\zeta_\alpha^\theta(0) = \alpha \in D_o$. Denote by $\tau_k(\alpha;\theta)$ the random time at which $\{\zeta_\alpha^\theta(t), t \geq 0\}$ is killed. Then

$$\text{Prob.}\{\&| \text{ the strategy } \theta \text{ is used and } \zeta_\alpha^\theta(0) = \alpha\}$$
$$= \int_C P_\alpha^\theta(\{\tau_k(\alpha;\theta) > \tau(\alpha,c;\theta)\} \cap \{\zeta_\alpha^\theta(\tau(\alpha,c;\theta)) \in K(c)\})\mu(dc). \tag{2.195}$$

Denote

$$V(\alpha,c;\theta) \triangleq P_\alpha^\theta(\{\tau_k(\alpha;\theta) > \tau(\alpha,c;\theta)\} \cap \{\zeta_\alpha^\theta(\tau(\alpha,c;\theta)) \in K(c)\}), \tag{2.196}$$

$$\alpha \in D_o, \quad c \in C, \quad \theta \in U.$$

where

$$U \triangleq \{\theta \in \tilde{U} : \sup_{c \in C} \sup_{\alpha \in D(c)} E_\alpha^\theta \, \tau(\alpha,c;\theta) < \infty\} .$$

In this example the following problem is considered: find a strategy $\theta \in U$ such that

$$\int_C V(\alpha,c;\theta^*)\mu(dc) \geq \int_C V(\alpha,c;\theta)\mu(dc) \text{ for any } \theta \in U \text{ and all } \alpha \in D_0. \quad (2.197)$$

A strategy $\theta^* \in U$ for which eqn (2.197) is satisfied will be called an *optimal strategy*.

Given $c \in C$, let $\mathcal{D}(c)$ denote the class of all functions $V = V(\alpha)$ such that: V is continuous on the closure $\bar{D}(c)$ of $D(c)$, and twice continuously differentiable on $D(c)$; for any $\theta \in U$, $\hat{\mathcal{L}}(\theta)V \in L_2(D(c))$, where

$$\begin{aligned}\hat{\mathcal{L}}(\theta)V(\alpha) \triangleq \sum_{i=1}^{2} \left(u_i(x) \frac{\partial V(\alpha)}{\partial x_i} + v_i(y) \frac{\partial V(\alpha)}{\partial y_i}\right) \\ + \tfrac{1}{2} \sum_{i=1}^{2} \left(\sigma_i^2 \frac{\partial^2 V(\alpha)}{\partial x_i^2} + \gamma_i^2 \frac{\partial^2 V(\alpha)}{\partial y_i^2}\right).\end{aligned} \quad (2.198)$$

Define, for $V \in \mathcal{D}(c)$,

$$\mathcal{L}(\theta)V(\alpha) = \hat{\mathcal{L}}(\theta)V(\alpha) - a(\alpha)V(\alpha), \quad \alpha \in D(c). \quad (2.199)$$

Lemma 2.4

Given $c \in C$, and $\theta \in U$. Let $V \in \mathcal{D}(c)$ satisfy

$$\mathcal{L}(\theta)V(\alpha) = 0, \ \alpha \in D(c); \quad (2.200)$$

$$V(\alpha) = 1, \ \alpha \in K(c); \quad V(\alpha) = 0, \ \alpha \notin D_0 ; \quad (2.201)$$

then

$$\begin{aligned}V(\alpha) &= P_\alpha^\theta(\{\tau_k(\alpha;\theta) > \tau(\alpha,c;\theta)\} \cap \{\zeta_\alpha^\theta(\tau(\alpha,c;\theta)) \in K(c)\}) \\ &= V(\alpha,c;\theta).\end{aligned} \quad (2.202)$$

Proof

From Theorem 9.7 of Dynkin ([81] , Vol.I,p.298) it follows that $\mathcal{L}(\theta)$ is the infinitesimal operator of $(\zeta_\alpha^\theta(t),P_\alpha^\theta)$. By applying Itô's formula (Gihman and Skorohod [1]) it can be shown that

$$E_\alpha^\theta V(\zeta_\alpha^\theta(\tau(\alpha,c;\theta))) = V(\alpha) + E_\alpha^\theta \int_0^{\tau(\alpha,c;\theta)} \mathcal{L}(\theta)V(\zeta_\alpha^\theta(s))ds, \qquad (2.203)$$

where E_α^θ denotes the expectation operation with respect to P_α^θ. Let V satisfy equations (2.200)-(2.201); then eqn (2.203) implies

$$\begin{aligned} V(\alpha) &= E_\alpha^\theta V(\zeta_\alpha^\theta(\tau(\alpha,c;\theta))) \\ &= \int_{\{\tau_k(\alpha;\theta) \, > \, \tau(\alpha,c;\theta)\}} V(\zeta_\alpha^\theta(\tau(\alpha,c;\theta)))(\omega)P_\alpha^\theta(d\omega) \\ &= \int_{\{\tau_k > \tau\}\cap\{\zeta_\alpha^\theta(\tau) \in K(c)\}} V(\zeta_\alpha^\theta(\tau))(\omega)P_\alpha^\theta(d\omega) \qquad (2.204) \\ &+ \int_{\{\tau_k > \tau\}\cap\{\zeta_\alpha^\theta(\tau) \in \partial D_0\}} V(\zeta_\alpha^\theta(\tau))(\omega)P_\alpha^\theta(d\omega) \\ &= \int_{\{\tau_k > \tau\}\cap\{\zeta_\alpha^\theta(\tau) \in K(c)\}} P_\alpha^\theta(d\omega) = V(\alpha,c;\theta) \end{aligned}$$

where $\tau_k = \tau_k(\alpha;\theta)$, $\tau = \tau(\alpha,c;\theta)$ and ∂D_0 denotes the boundary of D_0. □

Define

$$L(\theta) \triangleq \int_C \int_{D_0} (1 - V(\alpha,c;\theta))^2 d\alpha \, \mu(dc) \, , \, \theta \in U. \qquad (2.205)$$

A strategy $\theta^0 \in U$ for which $L(\theta^0) \le L(\theta)$ for all $\theta \in U$, will be called a *weak optimal strategy*.

In the same manner as in Section 2.6 we define

$$U_0(c) \triangleq \{\theta \in U : V(\cdot,c;\theta) \in \mathcal{D}(c) \text{ and satisfies (2.200)-(2.201)}\} \qquad (2.206)$$

and

$$U_0 \triangleq \bigcap_{c \in C} U_0(c). \qquad (2.207)$$

Given $c \in C$. We denote by $U_1(c)$ the class of all strategies $\theta \in U$ such that : there exists a sequence $\theta^{(n)} \in U_0(c)$, n=1,2,... which con= verges to θ, as $n \to \infty$, in the following sense:

(i) $V(\cdot,c;\theta^{(n)})$ converges (via (2.97)-(2.98)) to $V(\cdot,c;\theta) \in W^2(D_0)$ (where $V(\cdot,c;\theta^{(n)})$ here denotes the solution to (2.200)-(2.201) where $\theta = \theta^{(n)}$, n=1,2,...);

(ii)
$$\lim_{n\to\infty} \int_{D_0} |\mathcal{L}(\theta^{(n)})V(\alpha,c;\theta^{(n)}) + a(\alpha)I_{K(c)}(\alpha)|^4 d\alpha$$
$$= \int_{D_0} |\mathcal{L}(\theta)V(\alpha,c;\theta) + a(\alpha)I_{K(c)}(\alpha)|^4 \, d\alpha = 0, \tag{2.208}$$

where $\mathcal{L}(\theta)$ is given by (2.198)-(2.199).

(iii) $\sup_{\alpha\in D(c)} E_\alpha^\theta \tau(\alpha,c;\theta) \le M < \infty$

Now, the class U_1 is defined by

$$U_1 \triangleq \bigcap_{c\in C} U_1(c). \tag{2.209}$$

Take $H = \{V = V(\alpha,c) : \int_C \int_{D_0} V^2(\alpha,c)d\alpha\mu(dc) < \infty\}$. (2.210)

Let $V \in \mathcal{D}(c)$, $c \in C$, and define the following operators

$$LV(\alpha,c) \triangleq \tfrac{1}{2} \sum_{i=1}^{2} \left(\sigma_i^2 \frac{\partial^2 V(\alpha,c)}{\partial x_i^2} + \gamma_i^2 \frac{\partial^2 V(\alpha,c)}{\partial y_i^2}\right) - a(\alpha)V(\alpha,c); \tag{2.211}$$

$$L^*Q(\alpha,c) \triangleq \tfrac{1}{2} \sum_{i=1}^{2} \left(\sigma_i^2 \frac{\partial^2 Q(\alpha,c)}{\partial x_i^2} + \gamma_i^2 \frac{\partial^2 Q(\alpha,c)}{\partial y_i^2}\right) - a(\alpha)Q(\alpha,c), \tag{2.212}$$

for all functions Q such that $L^*Q \in L_2(D_0)$.

Since (2.100) is valid for any $c \in C$, then in the same manner as in Section 2.6, the following theorem can be obtained.

Theorem 2.5

Assume:

(a) the transition function $P_\theta(t,\alpha,\Gamma) = P_\alpha^\theta(\{\zeta_\alpha^\theta(t) \in \Gamma\}\cap\{\tau_k(\alpha;\theta) > t\})$, $t \ge 0$, $\alpha\in \mathbb{R}^4$, $\Gamma \in \mathcal{B}(\mathbb{R}^4)$, $\theta\in U_1$, has a density $p^\theta(t,x,z)$;

(b) $\{V_0(\cdot,c) : c \in C\}$, $\theta^0 \in U_1$, $\{Q_0(\cdot,c) : c \in C\}$ satisfy, for each $c \in C$,

$$\mathcal{L}(\theta^0)V_0(\alpha,c) = 0, \quad \text{a.e. in } D(c) \tag{2.213}$$

$$V_0(\alpha,c) = 1,\ \alpha \in K(c)\ ;\quad V_0(\alpha,c) = 0 \quad \alpha \notin D_0 \tag{2.214}$$

$$L^*Q_0(\alpha,c) = 1 - V_0(\alpha,c),\ \text{a.e. in } D_0 \tag{2.215}$$

$$Q_0(\alpha,c) = 0,\ \alpha \notin D_0 \tag{2.216}$$

where $\theta^0 \in U_1$ is determined by

$$\theta^0 = (u^0,v^0) = \arg\sup_{\theta \in U_1} \{- \sum_{i=1}^{2} \int_{D_x} u_i(x) \int_{D_y} \int_C Q_0(\alpha,c)(\partial V(\alpha,c;\theta)/\partial x_i \mu(dc)d\alpha$$

$$- \sum_{i=1}^{2} \int_{D_y} v_i(y) \int_{D_x} \int_C Q_0(\alpha,c)(\partial V(\alpha,c;\theta)/\partial y_i)\mu(dc)d\alpha\}\ ; \tag{2.217}$$

and $V(\cdot,c;\theta) \in W^2(D_0)$, $c \in C$, $\theta \in U_1$, satisfy

$$\mathcal{L}(\theta)V(\alpha,c;\theta) = 0, \quad \text{a.e. in } D(c) \tag{2.218}$$

$$V(\alpha,c;\theta) = 1,\ \alpha \in K(c);\ V(\alpha,c;\theta) = 0,\ \alpha \notin D_0. \tag{2.219}$$

Then

$$L(\theta^0) \le L(\theta) \text{ for any } \theta \in U_1. \tag{2.220}$$

2.10.3 Results

The procedure for computing weak suboptimal strategies, described in Section 2.7, here takes the following form.

Let $\mathbb{R}_h^4$ be a finite-difference grid on $\mathbb{R}^4$, with a constant mesh size h along all axes.

1. take $\theta^{(i)}(\alpha) = (u^{(i)}(x), v^{(i)}(y)) \in U_1$;

2. take $c^{(j)} \in C_h$ (where $C_h \triangleq C \cap \mathbb{R}_h^4$);

3. solve the following set of equations:

$$\mathcal{L}(\theta^{(i)})V(\alpha,c^{(j)}) = 0,\ \alpha \in D(c^{(j)});\ L^*Q(\alpha,c^{(j)}) = 1 - V(\alpha,c^{(j)}),\ \alpha \in D_0$$

$$V(\alpha,c^{(j)}) = Q(\alpha,c^{(j)}) = 0,\ \alpha \notin D_0;\ V(\alpha,c^{(j)}) = 1,\ \alpha \in K(c^{(j)});$$

4. put $j+1 \to j$ and go to 2;

5. after $V(\cdot,c^{(j)};\theta^{(i)})$ and $Q(\cdot,c^{(j)};\theta^{(i)})$ have been computed for all $c^{(j)} \in C_h$, go to 6;

6. compute $\theta^{(i+1)} \in U_1$ by

$$u_\ell^{(i+1)}(x) = -u_0 \operatorname{sign}\Big(\int_{D_y}\int_C Q(\alpha,c;\theta^{(i)})\frac{\partial V(\alpha,c;\theta^{(i)})}{\partial x_\ell}\mu(dc)dy\Big)\quad \ell=1,2;$$

$$v_\ell^{(i+1)}(y) = -v_0 \operatorname{sign}\Big(\int_{D_x}\int_C Q(\alpha,c;\theta^{(i)})\frac{\partial V(\alpha,c;\theta^{(i)})}{\partial y_\ell}\mu(dc)dx\Big)\quad \ell=1,2;$$

7. put $i+1 \to i$ and go to 1.

The procedure is stopped when $\theta^{(i+1)} = \theta^{(i)}$. The equations at stage 3 were here solved by using an upwind finite-difference method as described in Section 2.8.

Throughout the computations the following form for $\mu(dc)$ was taken:

$$\mu(dc) = \begin{cases} dc_1 dc_2/\pi R^2 & \text{for} \quad c \in \{c : c_1^2 + c_2^2 \le R^2\} \\ 0 & \text{for} \quad c \in \{c : c_1^2 + c_2^2 > R^2\}. \end{cases} \tag{2.221}$$

Also the function $a(\alpha)$ (eqn (2.179)) was taken as

$$a(\alpha) = a \quad \alpha \in D_0. \tag{2.222}$$

Computations were carried out for $\sigma_1^2 = \sigma_2^2 = \gamma_1^2 = \gamma_2^2 = 10^{-5},\ 10^{-6}$;

$\varepsilon = 0.1, 0.2, 0.25$; $u_o = v_o = 0.0, 0.05, 0.1, 0.15, \ldots, 1.0$;
$a = 0.0, 1/16, 2/16, 3/16, \ldots, 8/16$; $h = 0.25, 0.2$; and $R = 0.25, 0.0$.

In the case where $R = 0.25$ and $\varepsilon = 0.1$, the numerical procedure yielded two sequences, one of which was of the form $\theta^{(o)} = \theta^{(2)} = \theta^{(4)} = \ldots = \theta^{(2i)} = \tilde{\theta}_h$ for all $i > 0$, where

$$\tilde{\theta}_h = (-u_o \text{ sign } x_1, -u_o \text{ sign } x_2, -v_o \text{ sign } y_1, -v_o \text{ sign } y_2) \qquad (2.223)$$

and $\int_C V_o^h(\alpha,c;\tilde{\theta}_h)\mu(dc)$ is greater than $\int_C V_o^h(\alpha,c;\theta^{(2i+1)})\mu(dc)$ for all $i > 0$; while for $R = 0.25, 0.0$ and $\varepsilon = 0.2, 0.25$ the numerical proce= dure yielded a unique solution $\theta = \tilde{\theta}_h$, where $\tilde{\theta}_h$ is given by (2.223).

In Figs. 2.4 to 2.7 the following notations are used:

$$V(\alpha) \triangleq \int_C V_o^h(\alpha,c;\tilde{\theta}_h)\mu(dc) \qquad (2.224)$$

$$P = \int_C \int_{D_o} V_o^h(\alpha,c;\tilde{\theta}_h)d\alpha\mu(dc)/\int_{D_o} d\alpha. \qquad (2.225)$$

Fig. 2.4 shows the plots of P and $V(z^{(i)})$, i=1,2,3, as functions of $u_o = v_o$, for $R = 0.25$, $\sigma_1^2 = \sigma_2^2 = \gamma_1^2 = \gamma_2^2 = 10^{-5}$, $\varepsilon = 0.1$, $h = 0.25$ and $a = 2/8$; where $z^{(1)} = (0.75, -0.5, -0,75, -0.5)$, $z^{(2)} = (0.75,0.0,-0,75,0.0)$, $z^{(3)} = (0.0,0.25,0.0,-0.25)$.

Fig.2.5 shows the plots of P and $V(z^{(i)})$, i=1,2,3, as functions of $u_o = v_o$, for the same set of parameters as in Fig.2.4, but with $\varepsilon = 0.25$.

Fig.2.6 shows the plots of P and $V(\alpha^{(i)})$, i=1,2,3, as functions of $u_o = v_o$, for $R = 0.0$; $\sigma_1^2 = \sigma_2^2 = \gamma_1^2 = \gamma_2^2 = 10^{-6}$, $\varepsilon = 0.2$, $h = 0.2$ and $a = 1/16$; where $\alpha^{(1)} = (0.8, -0.6, -0.8, -0.6)$, $\alpha^{(2)} = (0.8,0.0,-0.8,0.0)$, $\alpha^{(3)} = (0.2, 0.0, -0.2, 0.0)$.

Fig.2.7 shows P and $V(\alpha^{(i)})$, i=1,2,3, as functions of a, for $R = 0.0$, $\sigma_1^2 = \sigma_2^2 = \gamma_1^2 = \gamma_2^2 = 10^{-6}$, $\varepsilon = 0.2$, $u_o = v_o = 0.4$ and $h = 0.2$.

The results displayed in Figs. 2.4 to 2.7, as well as other results, show that P and $V(\alpha)$ increase as $u_o = v_o$ increases, and that they decrease as a increases.

2.10.4 Conclusion

The numerical results obtained here sugget that $\tilde{\theta}_h$ is an optimal strategy.

In the case R = 0.0 (here c = 0, $0_o = 0_1 = 0_2$ and there are no navi= gational errors) this strategy implies that both M_1 and M_2 move continu= ally towards the origin ($0_o = 0_1 = 0_2$) and that the rendezvous will, with probability $V(\alpha,0;\tilde{\theta}_h)$, eventually take place near the origin.

In the case R = 0.25 (here c is random) the strategy $\tilde{\theta}_h$ implies that M_1 moves towards 0_1 and M_2 moves towards 0_2. Although the rendezvous will take place (with probability $\int_C V(\alpha,c;\tilde{\theta}_h)\mu(dc)$) it seems that a feed= back type strategy does not offer a complete solution to the rendezvous problem when c is random. If M_1 reaches 0_1 and M_2 reaches 0_2 without rendezvous having taken place, other policies have to be adopted, such as search strategies, for example. (For a comprehensive list of referen= ces on search theory see Strümpfer [98]).

The rendezvous problem treated here serves as an example of a problem of optimal control with partial observations, which can be solved only partially by feedback laws.

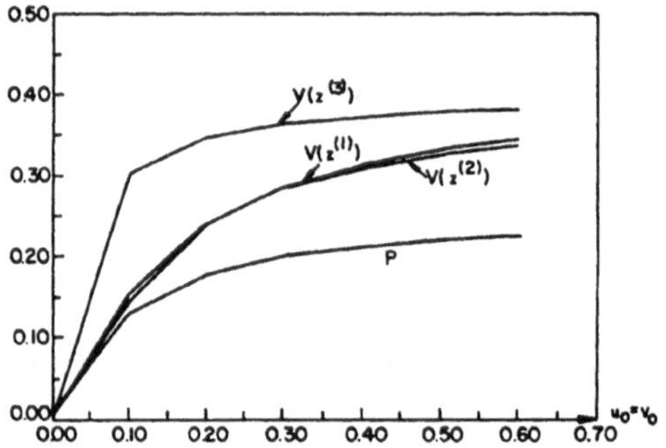

Fig. 2.4: The plots of P and $V(z^{(i)})$, i=1,2,3, as functions of $u_0 = v_0$, for R = 0.25, ε = 0.1, and a = 0.25.

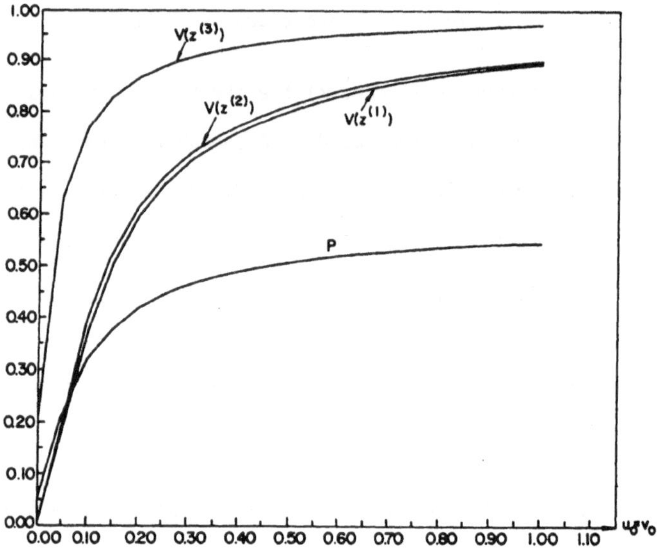

Fig.2.5: The plots of P and $V(z^{(i)})$, i=1,2,3, as functions of $u_0 = v_0$, for R = 0.25, ε = 0.25, and a = 0.25.

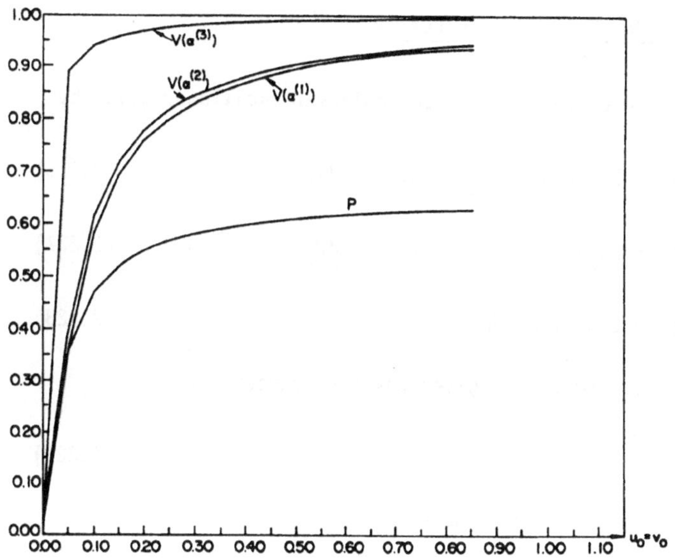

Fig.2.6: The plots of P and $V(\alpha^{(i)})$, i=1,2,3, as functions of $u_0 = v_0$, for R = 0.0, ε = 0.2, and a = 1/16.

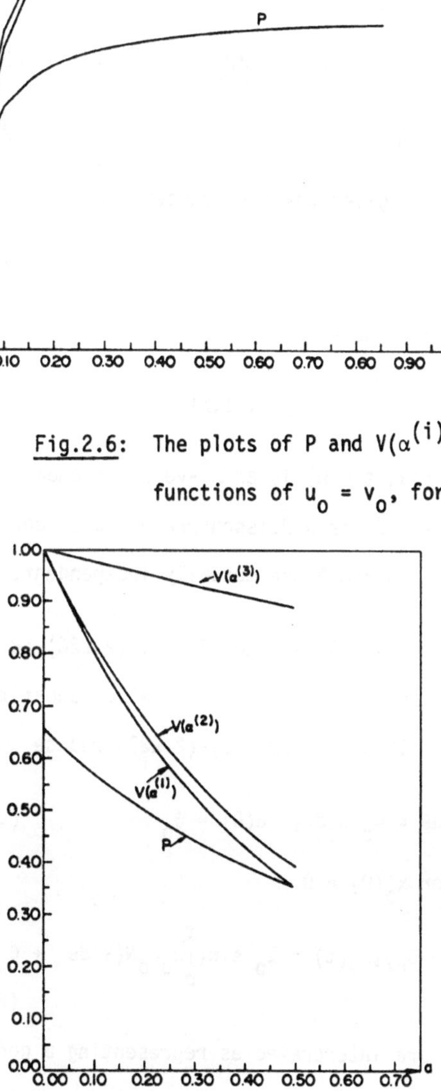

Fig.2.7: The plots of P and $V(\alpha^{(i)})$, i=1,2,3, as functions of a, for R = 0.0, ε = 0.2, and $u_0 = v_0 = 0.4$.

2.11 EXAMPLE 3 : CONTROL OF AN OSCILLATOR

Consider the noise-driven controlled nonlinear oscillator given by

$$dx_1 = [-w_0 x_2 x_3 + v_1(y)]dt + \sigma_1 dW_1 \tag{2.226}$$

$$dx_2 = [w_0 x_1 x_3 + v_2(y)]dt + \sigma_2\, dW_2 \tag{2.227}$$

$$dx_3 = c_0 dN + \sigma_3\, dW_3 \tag{2.228}$$

where w_0, σ_i, i=1,2,3, and c_0 are given positive numbers, and

$$y = (x_1, x_2). \tag{2.229}$$

$v : \mathbb{R}^2 \to U$ is the control vector, where

$$U \triangleq \{x \in \mathbb{R}^2 : |x_i| \le v_0,\ i=1,2\}. \tag{2.230}$$

$W \triangleq \{W(t) \triangleq (W_1(t), W_2(t), W_3(t)),\ t \ge 0\}$ is an $\mathbb{R}^3$-valued standard Wiener process and $N = \{N(t),\ t \ge 0\}$ is a Poisson process with constant parameter λ. It is assumed that W and N are mutually independent.

In order to illustrate the physical meaning of eqns (2.226)-(2.228) consider the case where $v_0 = 0$ and $\sigma_1 = \sigma_2 = \sigma_3 = 0$. Also, substitute $x_1 \triangleq Z \cos\theta$ and $x_2 \triangleq Z \sin\theta$. Then eqns (2.226)-(2.227) reduce to

$$dZ = 0 \quad Z(0) = Z_0; \quad d\theta = w_0\, x_3 dt, \quad \theta(0) = \theta_0 \tag{2.231}$$

from which it follows that, for $x_3(0) = 0$,

$$x_1(t) = Z_0 \cos(\int_0^t w_0 c_0 N(s)ds + \theta_0), x_2(t) = Z_0 \sin(\int_0^t w_0 c_0 N(s)ds + \theta_0) \tag{2.232}$$

Thus eqns (2.226)-(2.228) are interpreted as representing a nonlinear stochastic frequency-modulated sine wave oscillator. Let U be defined as in Section 2.5 (with τ_0 replacing τ). Given $v \in U$ and $x \in \mathbb{R}^3$, denote by $\zeta_x^v = \{\zeta_x^v(t) = (\zeta_{x,1}^v(t), \zeta_{x,2}^v(t), \zeta_{x,3}^v(t)),\ t \ge 0\}$ the (weak) solution to (2.226)-(2.228).

Denote by D_o and A the following sets in $\mathbb{R}^3$:

$$D_o \triangleq \{x : |x_i| < 1,\ i=1,2, \text{ and } -\varepsilon < x_3 < M_o\} \tag{2.233}$$

$$A \triangleq \{x : x \in D \text{ and } (\rho-\delta)^2 \le x_1^2 + x_2^2 \le (\rho+\delta)^2\} \tag{2.234}$$

where $0 < \varepsilon < < 1$, $0 < \delta < \rho$ and $M_o > 0$ are given numbers.

Let $\tau_o(x;v)$ denote the first exit time of ζ_x^v from D_o (eqn (2.5)) and define the following functional:

$$V(x;v) \triangleq E_x^v \Lambda \{t : 0 \le t < \tau_o(x;v),\ \zeta_x^v(t) \in A\} \tag{2.235}$$

where Λ is the Lebesgue measure on the real line.

In this example, the following optimal control problem is treated: find a strategy $v^* \in U$ such that

$$V(x;v^*) \ge V(x;v) \text{ for any } v \in U \text{ and all } x \in D_o. \tag{2.236}$$

Since this is a problem of the control of a partially observable stochastic system, the functional L(v) is here minimized on the class U_o defined in the sequel, where

$$L(v) \triangleq \int_{D_o} (T - V(x;v))^2 dx\ , \tag{2.237}$$

and it is assumed that T is given and satisfies

$$\sup_{x \in D} E_x^v \tau_o(x;v) < T < \infty \quad \text{for any } v \in U. \tag{2.238}$$

Let $\mathcal{D}_o$ denote the class of $V = V(x)$ as defined in Section 2.2 but with respect to the class U as defined in Section 2.5. Given $v \in U$. If $V \in \mathcal{D}_o$ satisfy

$$\mathcal{L}(v)V(x) = - I_A(x),\ x \in D_o;\ V(x) = 0,\ x \notin D_o; \tag{2.239}$$

then it can be shown that

$$V(x) = V(x;v) = E_x^v \Lambda\{t : 0 \le t < \tau_o(x;v),\ \zeta_x^v(t) \in A\},\ x \in D_o \tag{2.240}$$

Thus we here define

$$U_o \triangleq \{v \in U : V(\cdot;v) \in \mathcal{D}_o \text{ and satisfies } (2.239)\}.$$

Hence, by applying the same procedure as in the proof of Theorem 2.3, the following theorem is obtained.

Theorem 2.6

Suppose that $V_o \in \mathcal{D}_o$, $v^o \in U_o$ and Q_o satisfy

$$\mathcal{L}(v^o)V_o(x) = - I_A(x),\ x \in D_o \tag{2.241}$$

$$\mathcal{L}^* Q_o(x) = T - V_o(x),\quad \text{a.e. in } D_o \tag{2.242}$$

$$V_o(x) = Q_o(x) = 0,\ x \notin D_o \tag{2.243}$$

where $\mathcal{L}(v)$ and $\mathcal{L}^*$ are given by (2.245) and (2.246) respectively, and $v^o \in U_o$ is determined by

$$v^o = \underset{v \in U_o}{\operatorname{argsup}} \{- \sum_{i=1}^{2} \int_{D_{oy}} v_i(y) \int_{D_{o\tilde{x}}} Q_o(x)(\partial V(x;v)/\partial x_i) dx_3 dy\}, \tag{2.244}$$

and $D_{oy} = \{y = (x_1,x_2) : |x_i| < 1,\ i=1,2\}$, $D_{o\tilde{x}} = \{x_3 : -\varepsilon < x_3 < M_o\}$,

$$\mathcal{L}(v)V(x) = [-w_o x_2 x_3 + v_1(y)]\partial V(x)/\partial x_1 + [w_o x_1 x_3 + v_2(y)]\partial V(x)/\partial x_2$$
$$+ (\tfrac{1}{2}) \sum_{i=1}^{3} \sigma_i^2\, \partial^2 V(x)/\partial x_i^2 + \lambda[V(x_1,x_2,x_3+c_o) - V(x)] \tag{2.245}$$

$$\mathcal{L}^*Q(x) = w_o x_2 x_3 \; \partial Q(x)/\partial x_1 - w_o x_1 x_3 \; \partial Q(x)/\partial x_3$$
$$+ (\tfrac{1}{2}) \sum_{i=1}^{3} \sigma_i^2 \; \partial^2 Q(x)/\partial x_i^2 + \lambda[Q(x_1,x_2,x_3-c_o) - Q(x)]. \tag{2.246}$$

Then

$$L(v^o) \leq L(v) \text{ for any } v \in U_o. \tag{2.247}$$

The algorithm for computing weak suboptimal strategies, suggested in Section 2.7, together with the finite-difference procedure described in Section 2.8, have been applied for the following sets of parameters: $\sigma_1^2 = \sigma_2^2 = 10^{-5}$, $\sigma_3^2 = 10^{-15}$; $\delta = 0.05$; $\rho = 0.2, 0.3$; $w_o = 30,50,100,200,$ $500,1000,2000,5000,10000$; $c_o = 0.1$; $T = 100$; $\lambda = 1$; $v_o = 1,2,3$; $h = 0.1$ and $M_o = 1$. For all the cases computed, the sequence $\{v^{(n)}=(v_1^{(n)},v_2^{(n)})\}$ converged to $\bar{v}$ (always for $n \leq 7$) and the results indicate that

$$\bar{v}(y) = \begin{cases} (-v_o \operatorname{sign}(x_1), -v_o \operatorname{sign}(x_2)) & \text{if } x_1^2 + x_2^2 \geq \rho^2 \\ (v_o \operatorname{sign}(x_1), v_o \operatorname{sign}(x_2)) & \text{if } x_1^2 + x_2^2 < \rho^2 \end{cases} \tag{2.248}$$

Define

$$\ell^h(v) \triangleq h^3 \sum_{\substack{i,j,k \\ (ih,jh,kh) \in \bar{D}_{oh}}} (T - V^h(ih,jh,kh;v))^2 \tag{2.249}$$

where, for a given $h > 0$ and $v \in U_o$, $V^h(\cdot;v)$ denotes the solution to equations such as (2.144)-(2.145) obtained by applying a finite-dif= ference scheme to (2.239). Tables 2.4-2.5 show the values of $\{\ell^h(v^{(n)})\}$ and $\ell^h(\bar{v}^h)$ for some of the cases, where the $\{v^{(n)}\}$ are described in Section 2.7 and $\lim_{n\to\infty} v^{(n)} = \bar{v}^h$ on D_{oyh}, $D_{oyh} \triangleq \mathbb{R}_h^2 \cap D_{oy}$. The last element in each of the columns of $\ell^h(v^{(n)})$ is $\ell^h(\bar{v}^h)$.

Table 2.4: The values of $\{\ell^h(v^{(n)})\}$ and $\ell^h(\bar{v}^h)$ for $v_o = 1$, $\rho = 0.3$ and $w_o = 30,50,100$

n	$w_o = 30$ $\ell^h(v^{(n)})$	$w_o = 50$ $\ell^h(v^{(n)})$	$w_o = 100$ $\ell^h(v^{(n)})$
0	52754	52747	52837
1	52467	52606	52787
2	52125	52517	52757

Table 2.5: The values of $\{\ell^h(v^{(n)})\}$ and $\ell^h(\bar{v}^h)$ for $v_o = 2$, $\rho = 0.3$ and $w_o = 100,200,500,1000,2000,5000,10000$.

n	$w_o = 100$ $\ell^h(v^{(n)})$	$w_o = 200$ $\ell^h(v^{(n)})$	$w_o = 500$ $\ell^h(v^{(n)})$	$w_o = 1000$ $\ell^h(v^{(n)})$	$w_o = 2000$ $\ell^h(v^{(n)})$	$w_o = 5000$ $\ell^h(v^{(n)})$	$w_o = 10000$ $\ell^h(v^{(n)})$
0	52386	52737	52862	52881	52885	52887	52887
1	52386	52737	52862	52881	52885	52887	52887

Throughout the computation for the cases represented in Table 2.5 the strategy $v^{(o)}$ was chosen according to

$$v^{(o)}(y)=\begin{cases}(-v_o\,\text{sign}(x_1),\ -v_o\,\text{sign}(x_2)) & \text{for } x_1^2+x_2^2 \geq \rho^2 \\ (v_o\,\text{sign}(x_1),\ v_o\,\text{sign}(x_2)) & \text{for } x_1^2+x_2^2 < \rho^2\end{cases} \tag{2.250}$$

2.12 PROBABILISTIC INTERPRETATION OF (2.124)-(2.125)

Consider, for simplicity, the case where

$$\partial f_i(x)/\partial x_i = 0, \quad x \in \mathbb{R}^m, \quad i=1,\dots,m \tag{2.251}$$

$$\partial[(\sigma(x)\sigma'(x))_{ij}]/\partial x_i = 0, \quad x \in \mathbb{R}^m, \quad i,j=1,\dots,m \tag{2.252}$$

as happens to be the case in Examples 1-3 (Sections 2.9-2.11 respectively).

Then the operator $\mathcal{L}^*$ (eqn (2.102)) reduces to

$$\mathcal{L}^*Q(x) = - \sum_{i=1}^{m} f_i(x)\partial Q(x)/\partial x_i + (\tfrac{1}{2}) \sum_{i,j=1}^{m} (\sigma(x)\sigma'(x))_{ij} \, \partial^2 Q(x)/\partial x_i \partial x_j \tag{2.253}$$

$x \in D_o$

for any Q such that $\mathcal{L}^*Q \in L_2(D_o)$.

Consider the following nonlinear stochastic system:

$$dX = - f(X)dt + \sigma(X)d\tilde{W}, \quad t > 0, \quad X \in \mathbb{R}^m \tag{2.254}$$

where $\tilde{W} = \{\tilde{W}(t) = (\tilde{W}_1(t),\ldots,\tilde{W}_m(t)),\ t \geq 0\}$ is an $\mathbb{R}^m$-valued standard Wiener process on some probability space $(\tilde{\Omega},\tilde{F},\tilde{P})$, and f and σ satisfy the conditions stated in Section 2.5. Then (2.254) determines a diffusion process $\tilde{X}_x = \{\tilde{X}_x(t) = (\tilde{X}_{x,1}(t),\ldots,\tilde{X}_{x,m}(t)),\ t \geq 0\}$ such that $\tilde{X}_x(0) = x$. Furthermore, $\mathcal{L}^*$ is the infinitesimal generator of the family of strong Markov processes $\{(\tilde{X}_x,\tilde{P}_x),\ x \in \mathbb{R}^m\}$.

Now suppose that $Q_o \in W^2(D_o)$ satisfies, for $V_o \in W^2(D_o)$,

$$\mathcal{L}^*Q_o(x) = 1 - V_o(x), \quad \text{a.e. in } D_o \tag{2.124}$$

$$Q_o(x) = 0 \quad , \quad x \notin D_o \ , \tag{2.125}$$

and assume that:

(a) the transition function $\tilde{P}(t,x,\Gamma) = \tilde{P}_x(\{\tilde{X}_x(t) \in \Gamma\}),\ t \geq 0,\ x \in \mathbb{R}^m$, $\Gamma \in \mathcal{B}(\mathbb{R}^m)$, has a density $\tilde{p}(t,x,z)$;

(b) $\tilde{\tau}_o(x) < \infty$ $\tilde{P}_x$-almost surely, and $\tilde{E}_x\, \tilde{\tau}_o(x) \leq M < \infty$ for all $x \in D_o$, where $\tilde{\tau}_o(x)$ is the first exit time of $\tilde{X}_x$ from D_o (defined in the same manner as in (2.5)) and

$$\tilde{E}_x\, \tilde{\tau}_o(x) = \int_{\tilde{\Omega}} \tilde{\tau}_o(x)(\omega)\tilde{P}_x(d\omega) = \tilde{E}[\tilde{\tau}_o(x) \mid \tilde{X}_x(0) = x]. \tag{2.255}$$

Then, by using (2.100), it can be shown that

$$Q_o(x) = -\tilde{E}_x \int_0^{\tilde{\tau}_o(x)} [1 - V_o(\tilde{X}_x(t))]dt \ , \quad x \in \bar{D}_o. \qquad (2.256)$$

Under further restrictions, these results can be extended, to some classes of systems described by (2.1). We consider here only the case dealt with in Example 3 (Section 2.11).

The operator $\mathcal{L}^*$ given by (2.246) can be interpreted as the infinitesimal generator of a family of strong Markov processes $\{(\tilde{\tilde{X}}_x, \tilde{\tilde{P}}_x), x \in \mathbb{R}^3\}$ determined by the following set of stochastic differential equations:

$$dX_1 = w_o X_2 X_3 \, dt + \sigma_1 \, d\tilde{\tilde{W}}_1 \ , \quad t > 0 \qquad (2.257)$$

$$dX_2 = -w_o X_1 X_3 \, dt + \sigma_2 \, d\tilde{\tilde{W}}_2 \ , \quad t > 0 \qquad (2.258)$$

$$dX_3 = -c_o d\tilde{\tilde{N}} + \sigma_3 \, d\tilde{\tilde{W}}_3. \qquad (2.259)$$

Here $\tilde{\tilde{W}} = \{\tilde{\tilde{W}}(t) = (\tilde{\tilde{W}}_1(t), \tilde{\tilde{W}}_2(t), \tilde{\tilde{W}}_3(t)), \ t \geq 0\}$ and $\tilde{\tilde{N}} = \{\tilde{\tilde{N}}(t), \ t \geq 0\}$ are an $\mathbb{R}^3$-valued standard Wiener process and a Poisson process respectively, on a probability space $(\tilde{\tilde{\Omega}}, \tilde{\tilde{F}}, \tilde{\tilde{P}})$. It is assumed here that $\tilde{\tilde{W}}$ and $\tilde{\tilde{N}}$ are mutually independent.

Now suppose that $Q_o \in \mathcal{D}_o$ satisfies, for $V_o \in \mathcal{D}_o$,

$$\mathcal{L}^* Q_o(x) = T - V_o(x), \ x \in D_o; \ Q_o(x) = 0, \ x \notin D_o, \qquad (2.260)$$

then it can be shown, provided $\tilde{\tilde{E}}_x \, \tilde{\tilde{\tau}}_o(x) < \infty$, that

$$Q_o(x) = -\tilde{\tilde{E}}_x \int_0^{\tilde{\tilde{\tau}}_o(x)} (T - V_o(\tilde{\tilde{X}}_x(t)))dt \qquad (2.261)$$

where $\tilde{\tilde{\tau}}_o(x)$ is the first exit time of $\tilde{\tilde{X}}_x$ from D_o.

Since, in (2.256), $V_o(x) = P_x^{v^o}(\{\zeta_x^{v^o}(\tau(x;v^o)) \in K\}) \leq 1$, where $\zeta_x^{v^o}$ is determined by (2.105)-(2.106) with $v = v^o \in U_1$; and in (2.261), $V_o(x) = E_x^{v^o} \Lambda\{t : 0 \leq t < \tau_o(x;v^o), \ \zeta_x^{v^o}(t) \in A\} < T$, where $\zeta_x^{v^o}$ is determined by (2.226)-(2.228) with $v = v^o \in U_o$; it follows that $Q_o(x) \leq 0$,

$x \in D_0$, for the cases given by (2.256) and (2.261). The numerical re= sults obtained in Examples 1-3 (Sections 2.9-2.11 respectively) verified this characteristic.

CHAPTER 3

STRATEGIES USING INTERRUPTED OR SAMPLED OBSERVATIONS

3.1 INTRODUCTION

3.1.1 Interrupted Noisy Observations

Let (Ω,F,P) be a probability space. Consider the nonlinear stochastic system given by

$$dx = [f(x) + F(x)v(y)]dt + \sigma(x)dW \ , \ t > 0, \quad x \in \mathbb{R}^m \tag{3.1}$$

and let the observation process Y be determined by

$$dy = \theta x dt + \gamma(x)dB \quad , \ t > 0 \ , \ y \in \mathbb{R}^m \tag{3.2}$$

where $f : \mathbb{R}^m \to \mathbb{R}^m$, $F : \mathbb{R}^m \to \mathbb{R}^{m\times d}$, $\sigma : \mathbb{R}^m \to \mathbb{R}^{m\times m}$ and $\gamma : \mathbb{R}^m \to \mathbb{R}^{m\times m}$ are given functions; $v : \mathbb{R}^m \to U$, $U \subseteq \mathbb{R}^d$, is a feedback strategy.
$W \triangleq \{W(t) = (W_1(t),\dots,W_m(t)),\ t \geq 0\}$ and $B \triangleq \{B(t) = (B_1(t),\dots,B_m(t)),\ t \geq 0\}$ are two $\mathbb{R}^m$-valued standard Wiener processes on (Ω,F,P).
$\Theta \triangleq \{\theta(t),\ t \geq 0\}$ is a homogeneous jump Markov process on (Ω,F,P) with state space $S = \{0,1\}$ and transition probabilities

$$P(\theta(t+\Delta) = j \mid \theta(t) = i) = \begin{cases} q\Delta + o(\Delta) & \text{if } j\neq i \\ 1-q\Delta + o(\Delta) & \text{if } j=i \end{cases} \tag{3.3}$$

$i,j=o,1$

where $\pi_i = P(\theta(o)=i)$, $i=0,1$, and $q > 0$ are given. It is assumed that the processes W,B and Θ are mutually independent.

It is further assumed that f_i and F_{ij}, $i=1,\dots,m$, $j=1,\dots,d$ are bounded and continuously differentiable on $\mathbb{R}^m$ and that σ_{ij} and γ_{ij}, $i,j=1,\dots,m$ are bounded and twice continuously differentiable on $\mathbb{R}^m$.

Denote by $\tilde{U}$ the class of all strategies $v = v(y)$ such that v is bounded and continuously differentiable on $\mathbb{R}^m$. Let $v \in \tilde{U}$ and $\alpha = (x,y) \in \mathbb{R}^{2m}$. Then in the same manner as in Sergeeva and Teterina [99] and Sergeeva [100] it can be shown that equations (3.1)-(3.2) have a unique solution $\zeta_\alpha^v = \{\zeta_\alpha^v(t) = (\zeta_{\alpha,1}^v(t),\dots,\zeta_{\alpha,m}^v(t),\eta_{\alpha,1}^v(t),\dots,\eta_{\alpha,m}^v(t)),$ $t \geq 0\}$ which is such that $\zeta_\alpha^v(0) = \alpha$. Also, in the same manner as in Sergeeva and Teterina [99] it can be shown that (ζ_α^v,Θ) is a Markov pro= cess on (Ω,F,P). Furthermore, by following the same reasoning as in Kushner [101] (Section 5, Chapter 1) and using Theorem 3.10 of Dynkin [81] it can be shown that (ζ_α^v,Θ) is a strong Markov process. Note that the sample functions of ζ_α^v are continuous with probability 1.

Let D_o be an open and bounded set in $\mathbb{R}^{2m}$ and let K be a closed set, $K \subset D_o$. Denote $D \triangleq D_o - K$ and define

$$\tau_i(\alpha;v) \triangleq \begin{cases} \inf\{t : (\zeta_\alpha^v(t),\theta(t)) \in \partial D \times S \text{ when } (\zeta_\alpha^v(o),\theta(o))=(\alpha,i) \in D \times S\} \\ 0 \qquad \text{if } \zeta_\alpha^v(o) = \alpha \notin D \text{ and } \theta(o) = i \\ \infty \qquad \text{if } \zeta_\alpha^v(t) \in D \text{ for all } t \geq 0 \text{ and } \theta(o) = i \end{cases} \tag{3.4}$$

$i=0,1$, where ∂D denotes the boundary of D. Henceforward it is assumed in this work that $D_o = D_{ox} \times D_{oy}$, i.e. $\alpha = (x,y) \in D_o$ iff $x \in D_{ox}$ and $y \in D_{oy}$.

The following notations will be used in the sequel

$$P_{\alpha,i}(\cdot) \triangleq P(\cdot \mid (\zeta_\alpha^v(o),\theta(o)) = (\alpha,i)), \quad i=0,1 \tag{3.5}$$

$$E_{\alpha,i} \triangleq E[\cdot \mid (\zeta_\alpha^v(o),\theta(o)) = (\alpha,i)] , \quad i=0,1. \tag{3.6}$$

Denote

$$U \triangleq \{v \in \tilde{U} : \sup_{\alpha \in D} E_{\alpha,i}\, \tau_i(\alpha;v) < \infty , \ i=0,1\} \tag{3.7}$$

and define the following functionals

$$V_i(\alpha;v) \triangleq P_{\alpha,i}(\{\zeta_\alpha^v(\tau_i(\alpha;v)) \in K\}) - E_{\alpha,i} \int_0^{\tau_i(\alpha;v)} \sum_{j=1}^{d} \lambda_j(\theta(t)) v_j^2(\eta_\alpha^v(t))dt \tag{3.8}$$

$v \in U$, $i=0,1$,

and

$$V(\alpha;v) \triangleq \sum_{i=0}^{1} P(\theta(o) = i)V_i(\alpha;v) = \sum_{i=0}^{1} \pi_i \; V_i(\alpha;v) \tag{3.9}$$

where $\eta_\alpha^v(t) = (\eta_{\alpha,1}^v(t),\ldots,\eta_{\alpha,m}^v(t))$. $\lambda_j(\cdot)$, $j=1,\ldots,d$, are given func= tions satisfying $\lambda_j(0) \geq \lambda_j(1) \geq 0$, $j=1,\ldots,d$.

In Sections 3.2 and 3.3, necessary conditions for the maximization of $V(\cdot;v)$ on U, are established. In Section 3.4, sufficient conditions are derived on weak optimal strategies for the maximization of $V_i(\cdot;v)$, $i=0,1$, on a class U_o, in the case where $\lambda_j(\cdot) = 0$, $j=1,\ldots,d$.

3.1.2 Interrupted Observations

Let $(\tilde{\Omega},\tilde{F},\tilde{P})$ be a probability space. Consider the dynamical system represented by

$$dx = [f(x) + F(x)v(y)]dt + \sigma(x)dW \; , \; t > 0 \; , \; x \in \mathbb{R}^m \tag{3.1}$$

and let the observation process Y be given by

$$Y_t = \begin{cases} X_t = (x_1(t),\ldots,x_m(t)) & \text{if } \theta(t)=1 \\ \check{X}_t = (x_1(t),\ldots,x_p(t)) & \text{if } \theta(t)=0 \end{cases} \tag{3.10}$$

$0 < p < m$, where $f : \mathbb{R}^m \to \mathbb{R}^m$, $F : \mathbb{R}^m \to \mathbb{R}^{m\times d}$ and $\sigma : \mathbb{R}^m \to \mathbb{R}^{m\times m}$ are given functions satisfying the assumptions stated in subsection 3.1.1. $W \triangleq \{W(t) = (W_1(t),\ldots,W_m(t)), \; t \geq 0\}$ is an $\mathbb{R}^m$-valued standard Wiener process on $(\tilde{\Omega},\tilde{F},\tilde{P})$ and $\boldsymbol{\Theta} = \{\theta(t), \; t \geq 0\}$ is a homogeneous jump Markov process on $(\tilde{\Omega},\tilde{F},\tilde{P})$ with state space $S = \{0,1\}$ and transition probabilities $\{\tilde{P}(\theta(t+\Delta) = j|\theta(t)=i), \; i,j=0,1\}$ satisfying (3.3). $\pi_i = \tilde{P}(\theta(o)=i)$, $i=0,1$, and $q > 0$ are given numbers. It is assumed that the processes W and Θ are mutually independent. Owing to the nature of the observations, we take the strategy v to be of the form

$$v(Y_t) = \theta(t)u(X_t) + (1 - \theta(t))\check{v}(\check{X}_t) \ , \ t \geq 0. \tag{3.11}$$

It is tacitly assumed here that the process Θ is observable; i.e. the system knows at any $t \geq 0$ whether X_t or $\check{X}_t$ is observed.

Denote by $\tilde{U}$ the class of all strategies $v = (u,\check{v}) = \{(u(x),\check{v}(\check{x})), x \in \mathbb{R}^m, \check{x} \in \mathbb{R}^p\}$ such that $u : \mathbb{R}^m \to \mathcal{U}_1$ and $\check{v} : \mathbb{R}^p \to \mathcal{U}_o$, $\mathcal{U}_i \subseteq \mathbb{R}^d$ i=0,1, are bounded and continuously differentiable on $\mathbb{R}^m$ and $\mathbb{R}^p$ respectively. Let $v \in \tilde{U}$ and $x \in \mathbb{R}^m$. Then in the same manner as in Subsection 3.3.1 it can be shown that equations (3.1) and (3.11) have a unique solution $\zeta_x^v = \{\zeta_x^v(t) = (\zeta_{x,1}^v(t),\ldots,\zeta_{x,m}^v(t)), \ t \geq 0\}$ satisfying $\zeta_x^v(o) = x$, and that (ζ_x^v,Θ) is a strong Markov process. Again, note that the sample paths of ζ_x^v are continuous with probability 1.

Let D_o be an open and bounded set in $\mathbb{R}^m$ and let K be a closed set, $K \subset D_o$. Denote $D \triangleq D_o - K$ and define

$$\tau_i(x;v) \triangleq \begin{cases} \inf\{t: (\zeta_x^v(t),\theta(t)) \in \partial D \times S \text{ when } (\zeta_x^v(o),\theta(o))=(x,i) \in D \times S\} \\ 0 \qquad \text{if } \zeta_x^v(o) = x \notin D \text{ and } \theta(o) = i \\ \infty \qquad \text{if } \zeta_x^v(t) \in D \text{ for all } t \geq 0 \text{ and } \theta(o)=i \end{cases} \tag{3.12}$$

i=0,1.

Denote

$$P_{x,i}(\cdot) \triangleq \tilde{P}(\cdot \mid (\zeta_x^v(o),\theta(o)) = (x,i)), \quad i=0,1 \tag{3.13}$$

$$E_{x,i} \triangleq \tilde{E}[\ \cdot \mid (\zeta_x^v(o),\theta(o)) = (x,i)], \quad i=0,1 \tag{3.14}$$

where $\tilde{E}$ denotes the expectation operator with respect to $\tilde{P}$.
Define the following functionals

$$V_i(x;v) \triangleq P_{x,i}(\{\zeta_x^v(\tau_i(x;v)) \in K\}) \ , \ i=0,1 \ , \quad v \in U \tag{3.15}$$

$$V(x;v) \triangleq \sum_{i=0}^{1} \tilde{P}(\theta(0)=i)V_i(x;v) = \sum_{i=0}^{1} \pi_i \, V_i(x;v) \ , \quad v \in U \tag{3.16}$$

where

$$U \triangleq \{v = (u,\check{v}) \in \tilde{U} : \sup_{x \in D} E_{x,i} \, \tau_i(x;v) < \infty \ , \quad i=0,1\} \tag{3.17}$$

In Section 3.3, necessary conditions are derived for the maximization of $V(\cdot;v)$ on U.

3.2 NECESSARY CONDITIONS ON OPTIMAL STRATEGIES : $\mathcal{U} = \mathbb{R}^d$

In this section,without loss of generality, the system given by (3.1)-(3.2) is dealt with.

Let $\mathcal{D}$ denote the class of all pairs (V_0,V_1), $V_i : \mathbb{R}^{2m} \to \mathbb{R}$, $i=0,1$; such that V_i, $i=0,1$, are continuous on $\bar{D}_0$ ($D_0 \subset \mathbb{R}^{2m}$), twice continuously differentiable on D ($D \triangleq D_0 - K \subset \mathbb{R}^{2m}$), and such that $\partial V_i/\partial x_j$, $\partial V_i/\partial y_j$, $\partial^2 V_i/\partial x_\ell \, \partial x_k$ and $\partial^2 V_i/\partial y_\ell \, \partial y_k$ are in $L_2(D_0)$ for $i=0,1$, and $j,\ell,k=1,\dots,m$.

By using the same method as in Kushner [101] (Chapter 1, Section 5) for deriving the weak infinitesimal operator of (ζ_α^v,θ) (see also Krasovskii and Lidskii [33]) and using the fact that (ζ_α^v,θ) is a strong Markov pro= cess, the following equations are obtained

$$E_{\alpha,i} \, V_{\theta(\tau_i(\alpha;v))}(\zeta_\alpha^v(\tau_i(\alpha;v))) = V_i(\alpha) + E_{\alpha,i} \int_0^{\tau_i(\alpha;v)} \mathcal{L}_{\theta(t)}(v)(V_0(\zeta_\alpha^v(t)),V_1(\zeta_\alpha^v(t)))dt \tag{3.18}$$

$i=0,1$, $\alpha = (x,y)$, $v \in U$ (where U is given by (3.7))

where

$$\begin{aligned}\mathcal{L}_0(v)(V_0(\alpha),V_1(\alpha)) \triangleq {} & \sum_{i=1}^{m} [f_i(x) + \sum_{j=1}^{d} F_{ij}(x)v_j(y)]\partial V_0(\alpha)/\partial x_i \\ & + (\tfrac{1}{2}) \sum_{i,j=1}^{m} [\bar{\sigma}_{ij}(x)\partial^2 V_0(\alpha)/\partial x_i \partial x_j + \bar{\gamma}_{ij}(x)\partial^2 V_0(\alpha)/\partial y_i \partial y_j] \\ & - qV_0(\alpha) + qV_1(\alpha)\end{aligned} \tag{3.19}$$

and

$$\mathcal{L}_1(v)(V_0(\alpha),V_1(\alpha)) \triangleq \sum_{i=1}^{m} [f_i(x) + \sum_{j=1}^{d} F_{ij}(x)v_j(y)]\partial V_1(\alpha)/\partial x_i$$

$$+ \sum_{i=1}^{m} x_i\ \partial V_1(\alpha)/\partial y_i + (\tfrac{1}{2}) \sum_{i,j=1}^{m} [\bar{\sigma}_{ij}(x)\partial^2 V_1(\alpha)/\partial x_i\ \partial x_j \tag{3.20}$$

$$+ \bar{\gamma}_{ij}(x)\partial^2 V_1(\alpha)/\partial y_i\ \partial y_j] - qV_1(\alpha) + qV_0(\alpha)$$

$(V_0,V_1) \in \mathcal{D}$, $\alpha \in D_0$.

Here $\bar{\sigma}_{ij}(x) \triangleq (\sigma(x)\sigma'(x))_{ij}$ and $\bar{\gamma}_{ij}(x) \triangleq (\gamma(x)\gamma'(x))_{ij}$, $x \in \mathbb{R}^m$, $i,j=1,\ldots,m$.

Denote

$$\mathcal{L}(v)(V_0,V_1)(\alpha) \triangleq (\mathcal{L}_0(v)(V_0(\alpha),V_1(\alpha)),\ \mathcal{L}_1(v)(V_0(\alpha),V_1(\alpha)))$$

$$\triangleq (L_0(v)V_0(\alpha) + qV_1(\alpha),\ L_1(v)V_1(\alpha) + qV_0(\alpha)), \tag{3.21}$$

$(V_0,V_1) \in \mathcal{D}$, $v \in U$, $\alpha \in D_0$.

In the sequel the following lemma will be used. Here the notation $(a_1,b_1) = (a_2,b_2)$ stands for $a_1 = a_2$ and $b_1 = b_2$.

Lemma 3.1

Given $v \in U$. Let $(V_0,V_1) \in \mathcal{D}$ satisfy

$$\mathcal{L}(v)(V_0,V_1)(\alpha) = (\sum_{j=1}^{d} \lambda_j(0)v_j^2(y),\ \sum_{j=1}^{d} \lambda_j(1)v_j^2(y)),\ \alpha \in D \tag{3.22}$$

$$(V_0(\alpha),V_1(\alpha)) = (1,1),\ \alpha \in K\ ;\quad (V_0(\alpha),V_1(\alpha)) = (0,0)\ \alpha \notin D_0 \tag{3.23}$$

then

$$V_i(\alpha)=V_i(\alpha;v) = P_{\alpha,i}(\{\zeta_\alpha^v(\tau_i(\alpha;v)) \in K\}) - E_{\alpha,i} \int_0^{\tau_i(\alpha;v)} \sum_{j=1}^{d} \lambda_j(\theta(t))v_j^2(\eta_\alpha^v(t))dt \tag{3.24}$$

$i=0,1$, $\alpha \in \bar{D}_0$.

Proof

Let $(V_0,V_1) \in \mathcal{D}$ satisfy (3.22). Then (3.18) yields

$$V_i(\alpha)=E_{\alpha,i}V_{\theta(\tau_i(\alpha;v))}(\zeta_\alpha^v(\tau_i(\alpha;v)))-E_{\alpha,i}\int_0^{\tau_i(\alpha;v)}\sum_{j=1}^{d}\lambda_j(\theta(t))v_j^2(\eta_\alpha^v(t))dt \tag{3.25}$$

$i=0,1$.

Assume that (V_0,V_1) also satisfy (3.23). Then

$$E_{\alpha,i}V_{\theta(\tau_i(\alpha;v))}(\zeta_\alpha^v(\tau_i(\alpha;v)))$$

$$=\int_{\{\omega\,:\,\zeta_\alpha^v(\tau_i(\alpha;v))(\omega)\in K\}} V_{\theta(\tau_i(\alpha;v))(\omega)}(\zeta_\alpha^v(\tau_i(\alpha;v))(\omega))P_{\alpha,i}(d\omega) \tag{3.26}$$

$$+\int_{\{\omega\,:\,\zeta_\alpha^v(\tau_i(\alpha;v))(\omega)\notin D_0\}} V_{\theta(\tau_i(\alpha;v))(\omega)}(\zeta_\alpha^v(\tau_i(\alpha;v))(\omega))P_{\alpha,i}(d\omega)$$

$$=\int_{\{\omega\,:\,\zeta_\alpha^v(\tau_i(\alpha;v))(\omega)\in K\}} P_{\alpha,i}(d\omega)=P_{\alpha,i}(\{\zeta_\alpha^v(\tau_i(\alpha;v))\in K\})$$

which completes the proof of Lemma 3.1. □

Let $v\in U$. Define

$$J(v)\triangleq\int_{D_0}V(\alpha;v)d\alpha \tag{3.27}$$

where $V(\alpha;v)$ is given by (3.9). Suppose $v^*,v^0\in U$ are strategies satisfying $V(\alpha;v^*)\geq V(\alpha;v)$ for any $v\in U$ and all $\alpha\in D_0$ and $J(v^0)\geq J(v)$ for any $v\in U$. Then it can be shown that $J(v^0)=J(v^*)$ and consequently that $V(\alpha;v^0)=V(\alpha;v^*)$ a.e. in D_0. Hence a strategy $v^0\in U$ that maximizes $J(v)$ on U, whenever it exists, can be interpreted as an optimal strategy in some weak sense. In this section conditions are derived for the maximization of $J(v)$ on U in the case where $\mathcal{U}=\mathbb{R}^d$.

Theorem 3.1

Suppose there exists a strategy $v^0\in U$ such that

$$J(v^0)\geq J(v)\text{ for any }v\in U \tag{3.28}$$

Let $v^\varepsilon=v^0+\varepsilon\psi$, for all $\varepsilon\in[0,\varepsilon_0]$, $\varepsilon_0>0$, $\psi\in U$. Assume:

(i) for each $\varepsilon \in [0,\varepsilon_0]$ there is a unique element $(V_0^\varepsilon, V_1^\varepsilon) \in \mathcal{D}$ satisfying

$$\begin{cases} \mathcal{L}(v^\varepsilon)(V_0^\varepsilon, V_1^\varepsilon)(\alpha) = \left(\sum_{j=1}^{d} \lambda_j(o)(v_j^\varepsilon(y))^2, \sum_{j=1}^{d} \lambda_j(1)(v_j^\varepsilon(y))^2\right), \ \alpha \in D \\ (V_0^\varepsilon(\alpha), V_1^\varepsilon(\alpha)) = (1,1), \ \alpha \in K; \quad (V_0^\varepsilon(\alpha), V_1^\varepsilon(\alpha)) = (0,0), \ \alpha \notin D_0 \end{cases} \tag{3.29}$$

(ii) there is an element $(Q_0, Q_1) \in L_2(D_0) \times L_2(D_0)$ satisfying

$$\begin{cases} L_i^*(v^o)Q_i(\alpha) = -1 \ , \ i=0,1 \ , \quad \text{a.e. in } D_0 \\ Q_0(\alpha) = Q_1(\alpha) = 0 \ , \ \alpha \notin D_0 \end{cases} \tag{3.30}$$

(iii) $(V_i^\varepsilon - V_i^o)/\varepsilon$, $i=0,1$, converge weakly (in $L_2(D_0)$) as $\varepsilon \downarrow 0$ to 0;

(iv) $\partial V_i^\varepsilon/\partial x_j$, $i=0,1$, $j=1,\ldots,m$, converge weakly (in $L_2(D_0)$), as $\varepsilon \downarrow 0$ to $\partial V_i^o/\partial x_j$, $i=0,1$, $j=1,\ldots,m$, respectively

where

$$L_0^*(v)Q_0(\alpha) \triangleq - \sum_{i=1}^{m} \partial\{Q_0(\alpha)[f_i(x) + \sum_{j=1}^{d} F_{ij}(x)v_j(y)]\}/\partial x_i$$

$$+ (\tfrac{1}{2}) \sum_{i,j=1}^{m} \{\partial^2[Q_0(\alpha)\bar{\sigma}_{ij}(x)]/\partial x_i\, \partial x_j + \bar{\gamma}_{ij}(x)\partial^2 Q_0(\alpha)/\partial y_i\, \partial y_j\}$$

$$- q\, Q_0(\alpha) \tag{3.31}$$

and

$$L_1^*(v)Q_1(\alpha) \triangleq - \sum_{i=1}^{m} \partial\{Q_1(\alpha)[f_i(x) + \sum_{j=1}^{d} F_{ij}(x)v_j(y)]\}/\partial x_i - \sum_{i=1}^{m} x_i\, \partial Q_1(\alpha)/\partial y_i$$

$$+ (\tfrac{1}{2}) \sum_{i,j=1}^{m} \{\partial^2[Q_1(\alpha)\bar{\sigma}_{ij}(x)]/\partial x_i\, \partial x_j + \bar{\gamma}_{ij}(x)\partial^2 Q_1(\alpha)/\partial y_i\, \partial y_j\}$$

$$- q\, Q_1(\alpha). \tag{3.32}$$

Then

$$v_j^o(y) = \int_{D_{ox}} \sum_{i=1}^{m} F_{ij}(x)(\pi_o Q_o(\alpha)\partial V_o^o(\alpha)/\partial x_i$$

$$+ \pi_1 Q_1(\alpha)\partial V_1^o(\alpha)/\partial x_i)dx/2 \int_{D_{ox}} I_D(\alpha)(\pi_o\lambda_j(o)Q_o(\alpha) + \pi_1\lambda_j(1)Q_1(\alpha))dx$$

$$j=1,\dots,d, \quad y \in D_{oy}. \tag{3.33}$$

Proof

For each $\varepsilon \in [0,\varepsilon_o]$ let $(V_o^\varepsilon, V_1^\varepsilon)$ satisfy (3.29); then

$$L_o(v^o)(V_o^\varepsilon(\alpha)-V_o^o(\alpha))+q(V_1^\varepsilon(\alpha)-V_1^o(\alpha))+(L_o(v^\varepsilon)-L_o(v^o))\ V_o^\varepsilon(\alpha)$$
$$-k_o(v^\varepsilon(y))+k_o(v^o(y)) = 0, \quad \text{a.e. in } D_o \tag{3.34}$$

and

$$L_1(v^o)(V_1^\varepsilon(\alpha)-V_1^o(\alpha))+q(V_o^\varepsilon(\alpha)-V_o^o(\alpha))+(L_1(v^\varepsilon)-L_1(v^o))V_1^\varepsilon(\alpha)$$
$$-k_1(v^\varepsilon(y))+k_1(v^o(y)) = 0, \quad \text{a.e. in } D_o \tag{3.35}$$

where

$$k_i(v(y)) = \sum_{j=1}^{d} \lambda_j(i)v_j^2(y)I_D(\alpha) \quad , \quad i=0,1. \tag{3.36}$$

Multiplying (3.34) by $\pi_o Q_o$ and (3.35) by $\pi_1 Q_1$, adding the two ex= pressions and integrating their sum over D_o (and using the property $Q_o(\alpha) = Q_1(\alpha) = 0$, $\alpha \notin D_o$) yields

$$0 = \int_{D_o} \{(V_o^\varepsilon(\alpha)-V_o^o(\alpha))(\pi_o L_o^*(v^o)Q_o(\alpha)+\pi_1 qQ_1(\alpha))$$
$$+ (V_1^\varepsilon(\alpha)-V_1^o(\alpha))(\pi_1 L_1^*(v^o)Q_1(\alpha)+\pi_o qQ_o(\alpha))+\pi_o Q_o(\alpha)(-k_o(v^\varepsilon(y))+k_o(v^o(y)))$$
$$+ \pi_1 Q_1(\alpha)(-k_1(v^\varepsilon(y))+k_1(v^o(y))) \tag{3.37}$$
$$+ \sum_{i=1}^{m}\sum_{j=1}^{d} F_{ij}(x)(v_j^\varepsilon(y)-v_j^o(y))(\pi_o Q_o(\alpha)\partial V_o^\varepsilon(\alpha)/\partial x_i + \pi_1 Q_1(\alpha)\partial V_1^\varepsilon(\alpha)/\partial x_i)\}d\alpha$$

Now, by using (3,27) and (3.30) equation (3.37) reduces to

$$J(v^{\varepsilon}) - J(v^{o}) = \int_{D_o} [\pi_1 qQ_1(\alpha)(V_o^{\varepsilon}(\alpha)-V_o^{o}(\alpha))+\pi_o qQ_o(\alpha)(V_1^{\varepsilon}(\alpha)-V_1^{o}(\alpha))]d\alpha$$

$$+ \varepsilon \sum_{j=1}^{d} \int_{D_{oy}} \psi_j(y) \int_{D_{ox}} [\sum_{i=1}^{m} F_{ij}(x)(\pi_o Q_o(\alpha)\partial V_o^{\varepsilon}(\alpha)/\partial x_i$$

$$+ \pi_1 Q_1(\alpha)\partial V_1^{\varepsilon}(\alpha)/\partial x_i) - 2I_D(\alpha)v_j^{o}(y)[\pi_o \lambda_j(o)Q_o(\alpha) \tag{3.38}$$

$$+ \pi_1 \lambda_j(1)Q_1(\alpha)]dxdy$$

$$- \varepsilon^2 \sum_{j=1}^{d} \int_{D_o} I_D(\alpha)\psi_j^2(\alpha)(\pi_o \lambda_j(o)Q_o(\alpha)+\pi_1\lambda_j(1)Q_1(\alpha))d\alpha \le 0.$$

By using conditions (iii) and (iv) it follows that

$$\lim_{\varepsilon \to 0} (J(v^{\varepsilon})-J(v^{o}))/\varepsilon = \sum_{j=1}^{d} \int_{D_{oy}} \psi_j(y) \int_{D_{ox}} [\sum_{i=1}^{m} F_{ij}(x)(\pi_o Q_o(\alpha)\partial V_o^{o}(\alpha)/\partial x_i$$

$$+ \pi_1 Q_1(\alpha)\partial V_1^{o}(\alpha)/\partial x_i) - 2I_D(\alpha)v_j^{o}(y)[\pi_o \lambda_j(o)Q_o(\alpha) \tag{3.39}$$

$$+ \pi_1 \lambda_j(1)Q_1(\alpha)]dxdy \le 0$$

Hence v^{o} is given by (3.33). □

Condition (iii) of Theorem 3.1 can be dropped if eqns.(3.30) are replaced by another set of equations.

Theorem 3.2

Consider the case where $\pi_o > 0$ and $\pi_1 > 0$. Suppose there exists a strategy $v^{o} \in U$ such that

$$J(v^{o}) \ge J(v) \text{ for any } v \in U. \tag{3.28}$$

Let $v^{\varepsilon} = v^{o} + \varepsilon\psi$, for all $\varepsilon \in [0,\varepsilon_o]$, $\varepsilon_o > 0$, $\psi \in U$. Assume:

(i) for each $\varepsilon \in [0,\varepsilon_o]$ there is a unique element $(V_o^{\varepsilon},V_1^{\varepsilon}) \in \mathcal{D}$ satisfy= ing

$$\begin{cases} \mathcal{L}(v^{\varepsilon})(V_0^{\varepsilon},V_1^{\varepsilon})(\alpha) = (\sum_{j=1}^{d} \lambda_j(0)(v_j^{\varepsilon}(y))^2, \sum_{j=1}^{d} \lambda_j(1)(v_j^{\varepsilon}(y))^2), \ \alpha \in D \\ (V_0^{\varepsilon}(\alpha),V_1^{\varepsilon}(\alpha)) = (1,1), \ \alpha \in K; \quad (V_0^{\varepsilon}(\alpha),V_1^{\varepsilon}(\alpha)) = (0,0), \ \alpha \notin D_0 \end{cases} \tag{3.29}$$

(ii) there is an element $(Q_0,Q_1) \in L_2(D_0) \times L_2(D_0)$ satisfying

$$\begin{cases} L_0^*(v^0)Q_0(\alpha)+(\pi_1/\pi_0)qQ_1(\alpha) = -1, \quad \text{a.e. in } D_0 \\ L_1^*(v^0)Q_1(\alpha)+(\pi_0/\pi_1)qQ_0(\alpha) = -1, \quad \text{a.e. in } D_0 \\ Q_0(\alpha) = Q_1(\alpha) = 0, \ \alpha \notin D_0 \end{cases} \tag{3.40}$$

(iii) $\partial V_i^{\varepsilon}/\partial x_j$, i=0,1,, j=1,...,m, converge weakly (in $L_2(D_0)$), as $\varepsilon \downarrow 0$ to $\partial V_i^0/\partial x_j$, i=0,1, j=1,...,m, respectively.

Then v^0 is given by (3.33).

Proof

The proof of this theorem is identical to the proof of Theorem 3.1. □

Thus if one assumes that a strategy $v^0 \in U$ for which (3.28) is satis= fied, exists, and that all the conditions stated either in Theorem 3.1 or in Theorem 3.2 are satisfied, then in order to implement such a stra= tegy, the following system of equations has to be solved:

$$\begin{cases} L_0(v)V_0(\alpha) + qV_1(\alpha) = \sum_{j=1}^{d} \lambda_j(0)v_j^2(y) \quad , \ \alpha \in D \\ L_1(v)V_1(\alpha) + qV_0(\alpha) = \sum_{j=1}^{d} \lambda_j(1)v_j^2(y) \quad , \ \alpha \in D \\ V_0(\alpha) = V_1(\alpha) = 1, \ \alpha \in K; \quad V_0(\alpha) = V_1(\alpha) = 0, \ \alpha \notin D_0 \end{cases} \tag{3.41}$$

and either

$$\begin{cases} L_i^*(v)Q_i(\alpha) = -1, \ i=0,1, \quad \text{a.e. in } D_0 \\ Q_0(\alpha) = Q_1(\alpha) = 0, \ \alpha \notin D_0 \end{cases} \tag{3.42}$$

if the conditions of Theorem 3.1 are satisfied; or

$$L_0^*(v)Q_0(\alpha) + (\pi_1/\pi_0)qQ_1(\alpha) = -1, \quad \text{a.e. in } D_0$$

$$L_1^*(v)Q_1(\alpha) + (\pi_0/\pi_1)qQ_0(\alpha) = -1, \quad \text{a.e. in } D_0 \tag{3.43}$$

$$Q_0(\alpha) = Q_1(\alpha) = 0 \ , \ \alpha \notin D_0$$

if the conditions of Theorem 3.2 are satisfied; where

$$v_j(y) = \int_{D_{ox}} \sum_{i=1}^{m} F_{ij}(x)(\pi_0 Q_0(\alpha)\partial V_0(\alpha)/\partial x_i$$

$$+ \pi_1 Q_1(\alpha)\partial V_1(\alpha)/\partial x_i)dx/2 \int_{D_{ox}} I_D(\alpha)(\pi_0\lambda_j(o)Q_0(\alpha)+\pi_1\lambda_j(1)Q_1(\alpha))dx$$

$$j=1,\ldots,d \quad , y \in D_{oy}. \tag{3.44}$$

Results, similar to those given by Theorems 3.1 and 3.2, can be derived for the system given by (3.1),(3.10)-(3.11).

3.3 NECESSARY CONDITIONS ON OPTIMAL STRATEGIES : $\mathcal{U}_i$,i=0,1,ARE BOUNDED

In this section, without loss of generality, the system given by (3.1) and (3.10)-(3.11) is dealt with. Let

$$\mathcal{U}_0 \triangleq \{x \in \mathbb{R}^d : |x_i| \le \check{v}_{0,i}, \ i=1,\ldots,d\} \tag{3.45}$$

and

$$\mathcal{U}_1 \triangleq \{x \in \mathbb{R}^d : |x_i| \le u_{0,i}, \ i=1,\ldots,d\} \tag{3.46}$$

where $\check{v}_{0,i}$, i=1,...,d and $u_{0,i}$, i=1,...,d are given positive numbers. Denote by $\tilde{U}$ the class of all strategies $v = (u,\check{v}) = \{(u(x),\check{v}(\check{x})), x \in \mathbb{R}^m, \check{x} \in \mathbb{R}^p\}$ such that $u : \mathbb{R}^m \to \mathcal{U}_1$ and $\check{v} : \mathbb{R}^p \to \mathcal{U}_0$ are measurable.

Also, it is assumed that $f : \mathbb{R}^m \to \mathbb{R}^m$, $F : \mathbb{R}^m \to \mathbb{R}^{m\times d}$ and $\sigma : \mathbb{R}^m \to \mathbb{R}^{m\times m}$ satisfy the conditions stated in subsection 3.1.1. Since not all the ele= ments of $\tilde{U}$ can be precisely synthesized, or physically implemented, the following approach is adopted.

Define, for $v = (u,\check{v}) \in \tilde{U}$

$$u_a(x) \triangleq \int_{\mathbb{R}^m} u(x')\theta_a^{(m)}(x-x')dx' \tag{3.47}$$

$$\check{v}_a(\check{x}) \triangleq \int_{\mathbb{R}^p} \check{v}(\check{x}')\theta_a^{(p)}(\check{x}-\check{x}')d\check{x}' \tag{3.48}$$

where, for $0 < a << 1$, and $k=p,m$

$$\theta_a^{(k)}(x) \triangleq \begin{cases} \exp((|x/a|^2-1)^{-1})/h_a & |x|=|(x_1,\dots,x_k)|=[\sum_{i=1}^{k} x_i^2]^{\frac{1}{2}} < a \\ 0 & \text{for } |x| \geq a \end{cases} \tag{3.49}$$

and $h_a > 0$ is such that $\int_{\mathbb{R}^k} \theta_a^{(k)}(x)dx = 1$. The functions u_a and $\check{v}_a$ are continuously differentiable and $u_a(x) \in \mathcal{U}_1$, $\check{v}_a(\check{x}) \in \mathcal{U}_0$ for all $x \in \mathbb{R}^m$ and $\check{x} \in \mathbb{R}^p$ respectively. Henceforward in this section, instead of (3.1) and (3.10)-(3.11), the following equation is considered:

$$dx = [f(x) + \theta F(x)u_a(x) + (1-\theta)F(x)\check{v}_a(\check{x})]dt + \sigma(x)dW \tag{3.50}$$

$$t > 0, \quad x \in \mathbb{R}^m, \quad \check{x} \in \mathbb{R}^p.$$

Under the assumptions on f, F, σ and $\tilde{U}$; given $v \in \tilde{U}$ and $x \in \mathbb{R}^m$, equation (3.50) has a unique solution $\zeta_x^v = \{\zeta_x^v(t), t \geq 0\}$. Furthermore, $(\zeta_x^v,\boldsymbol{\theta})$ is a strong Markov process on $(\tilde{\Omega},\tilde{F},\tilde{P})$. We here use the notations ζ_x^v and $\tau_i(x;v)$, i=0,1, eq (3.12) (rather than $\zeta_x^{v_a}$ and $\tau_i(x;v_a)$, i=0,1 respectively) since the strategy $v = (u,\check{v})$ uniquely determines the stra= tegy $v_a = (u_a,\check{v}_a)$. We denote by U the class of strategies defined in (3.17), where D_o, K and D are defined as in subsection 3.1.2. Denote by $\mathcal{D}$ the class of all pairs (V_o,V_1), $V_i : \mathbb{R}^m \to \mathbb{R}$, i=0,1; such that V_i,i=0,1, are continuous on $\bar{D}_o$, twice continuously differentiable on D, and such that $\partial V_i/\partial x_j$ and $\partial^2 V_i/\partial x_\ell \, \partial x_k$, i=0,1, j,k,$\ell$=1,...,m, are in $L_2(D_o)$.

Define, as in (3.21)

$$\mathcal{L}(v)(V_0,V_1)(x) \triangleq (\mathcal{L}_0(\check{v})(V_0(x),V_1(x)),\ \mathcal{L}_1(u)(V_0(x),V_1(x)))$$

$$\triangleq (L_0(\check{v})V_0(x)+qV_1(x),\ L_1(u)V_1(x)+qV_0(x)) \tag{3.51}$$

$(V_0,V_1) \in \mathcal{D}$, $v = (u,\check{v}) \in U$, $x \in D_0$

where

$$L_0(\check{v})V_0(x) = \sum_{i=1}^{m} [f_i(x) + \sum_{j=1}^{d} F_{ij}(x)\check{v}_{a,j}(\check{x})]\partial V_0(x)/\partial x_i$$
$$+ (\tfrac{1}{2}) \sum_{i,j=1}^{m} (\sigma(x)\sigma'(x))_{ij}\ \partial^2 V_0(x)/\partial x_i\ \partial x_j - qV_0(x) \tag{3.52}$$

and

$$L_1(u)V_1(x) = \sum_{i=1}^{m} [f_i(x) + \sum_{j=1}^{d} F_{ij}(x)u_{a,j}(x)]\partial V_1(x)/\partial x_i$$
$$+ (\tfrac{1}{2}) \sum_{i,j=1}^{m} (\sigma(x)\sigma'(x))_{ij}\ \partial^2 V_1(x)/\partial x_i\ \partial x_j - qV_1(x). \tag{3.53}$$

Then, an equation like (3.18) can be obtained and the following lemma can be established.

Lemma 3.2

Given $v = (u,\check{v}) \in U$. Let $(V_0,V_1) \in \mathcal{D}$ satisfy

$$\mathcal{L}(v)(V_0,V_1)(x) = (0,0)\ ,\quad x \in D \tag{3.54}$$

$$(V_0(x),V_1(x)) = (1,1),\ x \in K;\quad (V_0(x),V_1(x)) = (0,0)\ ,\ x \notin D_0 \tag{3.55}$$

then

$$V_i(x) = V_i(x;v) = P_{x,i}(\{\zeta_x^v(\tau_i(x;v)) \in K\}),\ i=0,1,\quad x \in \bar{D}_0. \tag{3.56}$$

In this section necessary conditions are derived on a strategy $v^* \in U$ for which

$$J(v^*) \geq J(v) \quad \text{for any } v \in U \tag{3.57}$$

where

$$J(v) = \int_{D_o} V(x;v)dx = \int_{D_o} \sum_{i=o}^{1} \pi_i V_i(x;v)dx \tag{3.58}$$

and $V_i(x;v)$, $i=0,1$ and $V(x;v)$ are given by (3.15) and (3.16) respectively. Define

$$L_o^*(\check{v})Q_o(x) \triangleq - \sum_{i=1}^{m} \partial[(f_i(x) + \sum_{j=1}^{d} F_{ij}(x)\check{v}_{a,j}(\check{x}))Q_o(x)]/\partial x_i$$
$$+ (\tfrac{1}{2}) \sum_{i,j=1}^{m} \partial^2[(\sigma(x)\sigma'(x))_{ij}Q_o(x)]/\partial x_i\, \partial x_j - qQ_o(x) \tag{3.59}$$

$$L_1^*(u)Q_1(x) \triangleq - \sum_{i=1}^{m} \partial[(f_i(x) + \sum_{j=1}^{d} F_{ij}(x)u_{a,j}(x))Q_1(x)]/\partial x_i$$
$$+ (\tfrac{1}{2}) \sum_{i,j=1}^{m} \partial^2[(\sigma(x)\sigma'(x))_{ij}Q_1(x)]/\partial x_i\, \partial x_j - qQ_1(x) \tag{3.60}$$

for any (Q_o,Q_1) such that: for any $v = (u,\check{v}) \in U$, $L_o^*(\check{v})Q_o$ and $L_1^*(u)Q_1$ are in $L_2(D_o)$.

The following theorem gives necessary conditions on v^*.

Theorem 3.3

Consider the case where $\pi_o > 0$ and $\pi_1 > 0$. Suppose there exists a strategy $v^* = (u^*,\check{v}^*) \in U$ which satisfies (3.57).

Assume

(i) for any strategy $v = (u,\check{v}) \in U$ there exists a unique solution $(V_o(\cdot;v),V_1(\cdot;v)) \in \mathcal{D}$ to the equations

$$\mathcal{L}(v)(V_o,V_1)(x) = (0,0) \quad , \quad x \in D$$

$$(V_o(x),V_1(x)) = (1,1) \ , \ x \in K \ ; \quad (V_o(x),V_1(x)) = (0,0), \ x \notin D_o \tag{3.61}$$

where $\mathcal{L}(v)$ is defined by (3.51)-(3.53);

(ii) there is an element $(Q_o,Q_1) \in L_2(D_o) \times L_2(D_o)$ satisfying

$$\begin{aligned} &L_o^*(\check{v}^*)Q_o(x) + (\pi_1/\pi_o)qQ_1(x) = -1, \text{ a.e. in } D_o \\ &L_1^*(u^*)Q_1(x) + (\pi_o/\pi_1)qQ_o(x) = -1, \text{ a.e. in } D_o \\ &Q_o(x) = Q_1(x) = 0, \ x \notin D_o \end{aligned} \tag{3.62}$$

where L_o^* and L_1^* are defined by (3.59) and (3.60) respectively;

(iii) there is at least one strategy $v^{(1)} = (u^{(1)},\check{v}^{(1)}) \in U$ which satis= fies

$$\begin{aligned} &u_j(x) = u_{o,j} \text{ sign } \psi_{1,j}(x;v) \quad , \quad j=1,\dots,d \quad , \quad \text{a.e. in } D_o \\ &\check{v}_j(\check{x}) = \check{v}_{o,j} \text{ sign } \psi_{o,j}(\check{x};v) \quad , \quad j=1,\dots,d \quad , \quad \text{a.e. in } D_{o\check{x}} \end{aligned} \tag{3.63}$$

where

$$\begin{aligned} &\psi_{o,j}(\check{x};v) \triangleq \int_{D_{o\bar{x}}} \int_{D_{o\check{x}}} \theta_a^{(p)}(\check{x}-\check{x}') \sum_{i=1}^{m} \pi_o Q_o(\check{x}',\bar{x}) F_{ij}(\check{x}',\bar{x}) (\partial V_o(\check{x}',\bar{x};v)/\partial x_i) d\check{x}' d\bar{x} \\ &\qquad j=1,\dots,d \quad , \qquad \check{x} \in D_{o\check{x}} \\ &\psi_{1,j}(x;v) \triangleq \int_{D_o} \theta_a^{(m)}(x-x') \sum_{i=1}^{m} \pi_1 Q_1(x') F_{ij}(x') (\partial V_1(x';v)/\partial x_i) dx' \\ &\qquad j=1,\dots,d \quad , \ x \in D_o \end{aligned} \tag{3.64}$$

and

$$x = (\check{x},\bar{x}) \quad , \ \bar{x} = (x_{p+1},\dots,x_m) \quad , \ D_o = D_{o\check{x}} \times D_{o\bar{x}} \ .$$

Then

$$u_j^*(x) = u_{0,j} \text{ sign } \psi_{1,j}(x;v^*) \ , \ j=1,\dots,d \ , \text{ a.e. in } D_0$$
$$\check{v}_j^*(\check{x}) = \check{v}_{0,j} \text{ sign } \psi_{0,j}(\check{x};v^*) \ , \ j=1,\dots,d \ , \text{ a.e. in } D_{0\check{x}}. \tag{3.65}$$

(Without loss of generality, we here take $u(x) = 0$, $x \notin D_{0x}$ and $\check{v}(\check{x})=0$, $\check{x} \notin D_{0\check{x}}$, for any $v = (u,\check{v}) \in U$).

<u>Proof</u>

From (i) it follows that, for any $v = (u,\check{v}) \in U$

$$L_0(\check{v}^*)(V_0(x;v)-V_0(x;v^*))+q(V_1(x;v)-V_1(x;v^*))+(L_0(\check{v})-L_0(\check{v}^*))V_0(x;v)=0 \tag{3.66}$$

a.e. in D_0,

$$L_1(u^*)(V_1(x;v)-V_1(x;v^*))+q(V_0(x;v)-V_0(x;v^*))+(L_1(u)-L_1(u^*))V_1(x;v)=0 \tag{3.67}$$

a.e. in D_0.

Multiplying (3.66) by $\pi_0 Q_0$ and (3.67) by $\pi_1 Q_1$; adding the two ex= pressions; integrating their sum over D_0; using (ii) and performing some manipulations; we obtain

$$J(v)-J(v^*) = \sum_{j=1}^{d} \{\int_{D_{0\check{x}}} (\check{v}_j(\check{x})-\check{v}_j^*(\check{x}))\psi_{0,j}(\check{x};v)d\check{x}$$
$$+ \int_{D_0} (u_j(x)-u_j^*(x))\psi_{1,j}(x;v)dx\} \le 0. \tag{3.68}$$

By using (iii) the rest of the proof follows the same lines as the proof to Theorem 2.2. □

Hence, if it is assumed that a strategy $v^* = (u^*,\check{v}^*) \in U$ exists, for which (3.57) is satisfied, and that all the conditions stated in Theorem 3.3 are satisfied, then, in order to implement such a strategy, the following set of equations has to be solved:

$$\begin{cases} L_0(\check{v})V_0(x) + qV_1(x) = 0 \quad , x \in D \\ L_1(u)V_1(x) + qV_0(x) = 0 \quad , x \in D \\ V_0(x) = V_1(x) = 1,\ x \in K;\quad V_0(x) = V_1(x) = 0,\ x \notin D_0 \end{cases} \tag{3.69}$$

$$\begin{cases} L_0^*(\check{v})Q_0(x) + (\pi_1/\pi_0)qQ_1(x) = -1, \quad \text{a.e. in } D_0 \\ L_1^*(u\ )Q_1(x) + (\pi_0/\pi_1)qQ_0(x) = -1, \text{ a.e. in } D_0 \\ Q_0(x) = Q_1(x) = 0,\ x \notin D_0 \end{cases} \tag{3.70}$$

$$\begin{cases} u_j(x) = u_{o,j} \text{ sign } \psi_{1,j}(x;v) \ , \ j=1,\dots,d, \quad x \in D_0 \\ \check{v}_j(\check{x}) = \check{v}_{o,j} \text{ sign } \psi_{o,j}(\check{x};v) \ , \ j=1,\dots,d, \quad \check{x} \in D_{o\check{x}} \end{cases} \tag{3.71}$$

$$\begin{cases} u_a(x) = \int_{D_0} u(x')\theta_a^{(m)}(x-x')dx' \\ \check{v}_a(\check{x}) = \int_{D_{o\check{x}}} \check{v}(\check{x}')\theta_a^{(p)}(\check{x}-\check{x}')d\check{x}'. \end{cases} \tag{3.72}$$

Results, similar to that given by Theorem 3.3, can be derived for the system given by (3.1)-(3.2).

3.4 SUFFICIENT CONDITIONS ON WEAK OPTIMAL STRATEGIES

In this section, without loss of generality, the system given by (3.1)-(3.2) is considered. The definitions given here and the methods used can be easily adapted to the case where the system given by (3.1), (3.10) and (3.11) is considered.

It is assumed here that f_i and F_{ij}, i=1,...,m, j=1,...,d are bounded and continuously differentiable on $\mathbb{R}^m$ and that $\bar{\sigma}_{ij}$ and $\bar{\gamma}_{ij}$, i,j=1,...,m are bounded and twice continuously differentiable on $\mathbb{R}^m$.

Let U be given by (2.37) and denote by $\tilde{U}$ the class of all strategies

$v = \{v(y) : y \in \mathbb{R}^m\}$ such that $v : \mathbb{R}^m \to U$ is measurable. Also, in the same manner as in the previous section we define

$$v_a(y) = \int_{D_{oy}} v(y')\theta_a^{(m)}(y-y')dy' \tag{3.73}$$

where $\theta_a^{(m)}$ is given by (3.49) and where, without loss of generality, we take $v(y) = 0$, $y \notin D_{oy}$. Henceforward in this chapter, instead of equa= tions (3.1)-(3.2), the following system is considered:

$$dx = [f(x) + F(x)v_a(y)]dt + \sigma(x)dW,\ t > 0,\ x \in \mathbb{R}^m,\ v \in \tilde{U} \tag{3.74}$$

$$dy = \theta x dt + \gamma(x)dB,\ t > 0,\ y \in \mathbb{R}^m. \tag{3.75}$$

Using the same reasoning as in 3.1.1, it follows that equations (3.74)-(3.75) have a unique solution $\zeta_\alpha^v = \{\zeta_\alpha^v(t) = (\zeta_{\alpha,1}^v(t),\dots,\zeta_{\alpha,m}^v(t),\eta_{\alpha,1}^v(t),\dots,\eta_{\alpha,m}^v(t)),\ t \geq 0\}$, $v \in \tilde{U}$, $\alpha = (x,y) \in \mathbb{R}^{2m}$, and that (ζ_α^v,Θ) is a strong Markov process on (Ω,F,P). The sets D_0, D and K are taken here to be the same as in subsection 3.1.1 and the class $\mathcal{D}$ is defined in the same manner as in Section 3.2. The random times $\tau_i(\alpha;v)$, $i=0,1$, and the class U are defined as in subsection 3.1.1 (eqns (3.4) and (3.7) respec= tively, but where ζ_α^v is determined by (3.74)-(3.75)). Here we denote

$$V_i(\alpha;v) \triangleq P_{\alpha,i}(\{\zeta_\alpha^v(\tau_i(\alpha;v)) \in K\}),\ i=0,1,\ v \in U,\ \alpha \in \bar{D}_0. \tag{3.76}$$

Define

$$\ell(v) \triangleq \sum_{i=o}^{1} \int_{D_0} (1 - V_i(\alpha;v))^2 d\alpha \quad ,\ v \in U. \tag{3.77}$$

Let $v^* \in U$ satisfy $V_i(\alpha;v^*) \geq V_i(\alpha;v)$, $i=0,1$, for any $v \in U$ and all $\alpha \in D_0$, and let $v^o \in U$ satisfy $\ell(v^o) \leq \ell(v)$ for any $v \in U$. Then it can be shown that $V_i(\alpha;v^*) = V_i(\alpha;v^o)$ a.e. in D_0, $i=0,1$. This property of v^o leads us to the notion of weak optimal strategy.

A strategy $v^o \in U$ for which $\ell(v^o) \leq \ell(v)$ for any $v \in U$ will here be called a *weak optimal strategy*. Although $\ell(v)$ is properly defined on U, we are here interested in the minimization of $\ell(v)$ on a subclass $U_0 \subseteq U$ to be defined below.

Consider the equations

$$\begin{cases} \mathcal{L}(v)(V_0,V_1)(\alpha) = (0,0) \quad, \alpha \in D \quad, \quad v \in U \\ \\ (V_0(\alpha),V_1(\alpha)) = (1,1) \quad, \alpha \in K; \quad (V_0(\alpha),V_1(\alpha)) = (0,0), \ \alpha \notin D_0 \end{cases} \tag{3.78}$$

where

$$\mathcal{L}(v)(V_0,V_1)(\alpha) \triangleq (\mathcal{L}_0(v_a)(V_0(\alpha),V_1(\alpha)),\ \mathcal{L}_1(v_a)(V_0(\alpha),V_1(\alpha))) \tag{3.79}$$

and $\mathcal{L}_i(v_a)(V_0(\alpha),V_1(\alpha))$, i=0,1, are given by (3.19) and (3.20) respectively, with $v=v_a$ inserted there. In the same manner as in (2.85) we define

$$U_0 \triangleq \{v \in U : (V_0(\cdot;v),V_1(\cdot;v)) \in \mathcal{D} \text{ and satisfies (3.78)}\} \subseteq U . \tag{3.80}$$

Note that if $(V_0,V_1) \in \mathcal{D}$ satisfies equations (3.78) then

$$V_i(\alpha) = V_i(\alpha;v) = P_{\alpha,i}(\{\zeta_\alpha^v(\tau_i(\alpha;v)) \in K\}),\ i=0,1,\ \alpha \in \bar{D}_0 .$$

In this section sufficient conditions are derived for the minimization of $\ell(v)$ on U_0. This is done via the use of Lemma 2.1.

Consider the set of ordered pairs (V_0,V_1), $V_i \in L_2(D_0)$, i=0,1. A vector space is defined on this set by means of the equations

$$\lambda(V_0,V_1) = (\lambda V_0,\lambda V_1) \quad \text{for any } \lambda \in \mathbb{R}$$

$$(V_0,V_1) + (W_0,W_1) = (V_0 + W_0,V_1 + W_1),\ W_i \in L_2(D_0),\ i=0,1 \tag{3.81}$$

and the zero element is (o,o). (Here o stands for the zero element in $L_2(D_0)$).

With the scalar product defined by:

$$< (V_0,V_1),(W_0,W_1) > \triangleq \sum_{i=o}^{1} \int_{D_0} V_i(\alpha)W_i(\alpha)d\alpha \tag{3.82}$$

this space becomes a Hilbert space that will be denoted by H.

Let $v \in U_0$, then $\ell(v)$ can be written as

$$\ell(v) = < (1-V_0(\cdot;v),1-V_1(\cdot;v)),\ (1-V_0(\cdot;v),1-V_1(\cdot;v)) > \tag{3.83}$$

and

$$\mathcal{L}(v)(V_0(\cdot;v),V_1(\cdot;v))(\alpha) = (0,0)\ ,\ \text{a.e. in } D_0. \tag{3.84}$$

Let $(V_0,V_1) \in \mathcal{D}$. Define the following operator on $\mathcal{D}$:

$$\mathcal{L}(V_0,V_1) \triangleq (\mathcal{L}_0 V_0 + qV_1,\ \mathcal{L}_1 V_1 + qV_0) \tag{3.85}$$

where

$$\mathcal{L}_0\, V_0(\alpha) \triangleq \sum_{i=1}^{m} f_i(x)\partial V_0(\alpha)/\partial x_i + (\tfrac{1}{2}) \sum_{i,j=1}^{m} [\bar{\sigma}_{ij}(x)\partial^2 V_0(\alpha)/\partial x_i\, \partial x_j$$
$$+ \bar{\gamma}_{ij}(x)\partial^2 V_0(\alpha)/\partial y_i\, \partial y_j)] - qV_0(\alpha)\ ,\ \alpha \in D_0 \tag{3.86}$$

and

$$\mathcal{L}_1 V_1(\alpha) \triangleq \sum_{i=1}^{m} (f_i(x)\partial V_1(\alpha)/\partial x_i + x_i \partial V_1(\alpha)/\partial y_i)$$
$$+ (\tfrac{1}{2}) \sum_{i,j=1}^{m} [\bar{\sigma}_{ij}(x)\partial^2 V_1(\alpha)/\partial x_i\, \partial x_j + \bar{\gamma}_{ij}(x)\partial^2 V_1(\alpha)/\partial y_i\, \partial y_j] \tag{3.87}$$
$$- qV_1(\alpha)\ \ ,\ \alpha \in D_0.$$

Also, define the following element $\psi = (\psi_0,\psi_1) \in H$ by

$$\psi_0 \triangleq \mathcal{L}_0^*\, Q_0 + qQ_1 \quad ,\quad \psi_1 \triangleq \mathcal{L}_1^* Q_1 + qQ_0 \tag{3.88}$$

where $\mathcal{L}_i^*$ is the adjoint operator of $\mathcal{L}_i$, i=0,1, and $(Q_0,Q_1) \in H$ is such that $\mathcal{L}_i^* Q_i \in L_2(D_0)$, i=0,1.

Suppose $(Q_o,Q_1) \in H$ is such that

$$\mathcal{L}_i^* Q_i \in L_2(D_o) \;,\; i=0,1 \quad \text{and} \quad Q_i(\alpha) = 0,\; \alpha \notin D_o,\; i=0,1 \tag{3.89}$$

then by using (3.89), (3.88) and (3.84) where $v \in U_o$, we obtain

$$< (\psi_o,\psi_1),(V_o(\cdot;v),V_1(\cdot;v)) > = \sum_{i=o}^{1} \int_{D_o} \psi_i(\alpha)V_i(\alpha;v)d\alpha$$

$$= \int_{D_o} V_o(\alpha;v)[\mathcal{L}_o^* Q_o(\alpha) + qQ_1(\alpha)]d\alpha + \int_{D_o} V_1(\alpha;v)[\mathcal{L}_1^* Q_1(\alpha) + qQ_o(\alpha)]d\alpha$$

$$= \int_{D_o} Q_o(\alpha)[\mathcal{L}_o V_o(\alpha;v) + qV_1(\alpha;v)]d\alpha + \int_{D_o} Q_1(\alpha)[\mathcal{L}_1 V_1(\alpha;v) + qV_o(\alpha;v)]d\alpha \tag{3.90}$$

$$= - \sum_{j=1}^{d} \int_{D_{oy}} v_{a,j}(y) \int_{D_{ox}} \sum_{\ell=o}^{1} Q_\ell(\alpha) \sum_{i=1}^{m} F_{ij}(x)(\partial V_\ell(\alpha;v)/\partial x_i)dxdy$$

$$= - \sum_{j=1}^{d} \int_{D_{oy}} v_j(y')\Lambda_j(y';v)dy'$$

where

$$\Lambda_j(y;v) \triangleq \int_{D_{oy}} \theta_a^{(m)}(y-y') \int_{D_{ox}} \sum_{\ell=o}^{1} Q_\ell(x,y') \sum_{i=1}^{m} F_{ij}(x)(\partial V_\ell(x,y';v)/\partial x_i)dxdy' \tag{3.91}$$

$j=1,\ldots,d \quad , y \in D_{oy}$.

In order to make use of Lemma 2.1 let

$$A \triangleq \{(V_o(\cdot;v),V_1(\cdot;v)) : v \in U_o\} \tag{3.92}$$

then (2.75) and (3.90) yield

$$\sup_{(V_o(\cdot;v),V_1(\cdot;v)) \in A} < (\psi_o,\psi_1),(V_o(\cdot;v),V_1(\cdot;v)) >$$

$$= \sup_{v \in U_o} \{- \sum_{j=1}^{d} \int_{D_{oy}} v_j(y)\Lambda_j(y;v)dy\}. \tag{3.93}$$

In order to satisfy (2.76) it is sufficient to take $V_d(\alpha) = (1,1)$ for all $\alpha \in D_o$, and to choose (Q_o,Q_1) such that

$$\mathcal{L}_o^* Q_o(\alpha) + qQ_1(\alpha) = 1 - V_o(\alpha;v^o) \quad , \text{a.e. in } D_o \tag{3.94}$$

$$\mathcal{L}_1^* Q_1(\alpha) + qQ_o(\alpha) = 1 - V_1(\alpha;v^o) \quad , \text{a.e. in } D_o \tag{3.95}$$

$$Q_o(\alpha) = Q_1(\alpha) = 0 \quad , \quad \alpha \notin D_o \tag{3.96}$$

where v^o is determined by

$$v^o = \arg\sup_{v \in U_o} \{- \sum_{j=1}^{d} \int_{D_{oy}} v_j(y)\Lambda_j(y;v)dy\}. \tag{3.97}$$

The following theorem is a straightforward conclusion of this section.

Theorem 3.4

Suppose $(V_o(\cdot;v^o),V_1(\cdot;v^o))$, $v^o \in U_o$ and (Q_o,Q_1) satisfy

$$\mathcal{L}_i(v_a^o)(V_o(\alpha;v^o),V_1(\alpha;v^o)) = 0 \ , \ \alpha \in D \quad , \ i=0,1 \tag{3.98}$$

$$\begin{cases} \mathcal{L}_o^* Q_o(\alpha) + qQ_1(\alpha) = 1 - V_o(\alpha;v^o), \text{ a.e. in } D_o \\ \mathcal{L}_1^* Q_1(\alpha) + qQ_o(\alpha) = 1 - V_1(\alpha;v^o), \text{ a.e. in } D_o \end{cases} \tag{3.99}$$

$$\begin{cases} (V_o(\alpha;v^o),V_1(\alpha;v^o)) = (1,1) \quad , \ \alpha \in K, \\ (V_o(\alpha;v^o),V_1(\alpha;v^o)) = (Q_o(\alpha),Q_1(\alpha)) = (0,0), \ \alpha \notin D_o \end{cases} \tag{3.100}$$

where $v^o \in U_o$ is determined by

$$v^o = \arg\sup_{v \in U_o} \{ \sum_{j=1}^{d} \int_{D_{oy}} v_j(y)\Lambda_j(y;v)dy\} \tag{3.101}$$

and

$$v_a^o(y) = \int_{D_{oy}} v^o(y')\theta_a^{(m)}(y-y')dy' \quad ; \tag{3.102}$$

then

$$\ell(v^o) \le \ell(v) \quad \text{for any } v \in U_o. \tag{3.103}$$

$(\mathcal{L}_i(v_a^o)$, i=0,1, are given by (3.19) and (3.20) respectively, where $v = v_a^o$ there, and Λ_j, j=1,...,d are given by (3.91)).

Theorem 3.4 states sufficient conditions for the minimization of $\ell(v)$ on U_o.

Remark 3.1

Consider the system given by (3.1) and (3.10)-(3.11), or given by (3.50). Assume f, F and σ satisfy the assumptions stated in subsection 3.1.1, and let D_o, K, D, $\tau_i(x;v)$, i=0,1; $\mathcal{U}_i$, i=0,1; u_a, $\check{v}_a$, U and $\mathcal{D}$ be as described in Section 3.3. Also, define the class

$$U_o \triangleq \{v = (u,\check{v}) : (V_o(\cdot;v),V_1(\cdot;v)) \in \mathcal{D} \text{ and satisfies (3.54) and (3.55)}\} ; \tag{3.104}$$

the functional

$$\ell(v) \triangleq \sum_{i=o}^{1} \int_{D_o} (1 - V_i(x;v))^2 dx, \quad v \in U; \tag{3.105}$$

and the operators

$$\mathcal{L}_k^* Q_k(x) \triangleq - \sum_{i=1}^{m} \partial[f_i(x)Q_k(x)]/\partial x_i + (\tfrac{1}{2}) \sum_{i,j=1}^{m} \partial^2[(\sigma(x)\sigma'(x))_{ij}Q_k(x)]/\partial x_i \, \partial x_j - qQ_k(x) \tag{3.106}$$

k=0,1, $x \in D_o$

for any (Q_o,Q_1) such that $\mathcal{L}_k^* Q_k \in L_2(D_o)$, k=0,1.

By using the same techniques that led to the establishment of Theorem 3.4, the following theorem is obtained.

Theorem 3.5

Suppose $(V_o(\cdot;v^o),V_1(\cdot;v^o))$, $v^o \in U_o$ and (Q_o,Q_1) satisfy

$$\mathcal{L}(v^o)(V_o,V_1)(x) = (0,0) \quad , x \in D \tag{3.107}$$

(where $\mathcal{L}(v^o)$ is given by (3.51)-(3.53), in which $\check{v}_a = \check{v}_a^o$ is inserted in (3.52) and $u_a = u_a^o$ is inserted in (3.53), and

$$u_a^o(x) = \int_{D_o} u^o(x')\theta_a^{(m)}(x-x')dx', \quad \check{v}_a^o(\check{x}) = \int_{D_{o\check{x}}} \check{v}^o(\check{x}')\theta_a^{(p)}(\check{x}-\check{x}')d\check{x}');$$

$$\begin{cases} \mathcal{L}_o^* Q_o(x) + qQ_1(x) = 1 - V_o(x;v^o), \text{ a.e. in } D_o \\ \mathcal{L}_1^* Q_1(x) + qQ_o(x) = 1 - V_1(x;v^o), \text{ a.e. in } D_o \end{cases} \tag{3.108}$$

$$\begin{cases} (V_o(x;v^o), V_1(x;v^o)) = (1,1), \ x \in K \\ (V_o(x;v^o), V_1(x;v^o)) = (Q_o(x), Q_1(x)) = (0,0), \ x \notin D_o \end{cases} \tag{3.109}$$

where $v^o = (u^o, \check{v}^o) \in U_o$ is determined by

$$\begin{aligned}(u^o, \check{v}^o) = \arg \sup_{(u,\check{v}) \in U_o} \{ &- \sum_{j=1}^{d} \int_{D_{o\check{x}}} \check{v}_j(\check{x}')\Lambda_{o,j}(\check{x}';v)d\check{x}' \\ &- \sum_{j=1}^{d} \int_{D_o} u_j(x')\Lambda_{1,j}(x';v)dx' \}\end{aligned} \tag{3.110}$$

and

$$\Lambda_{o,j}(\check{x};v) \triangleq \int_{D_{o\check{x}}} \theta_a^{(p)}(\check{x}-\check{x}') \int_{D_{o\bar{x}}} Q_o(\check{x}',\bar{x}) \sum_{i=1}^{m} F_{ij}(\check{x}',\bar{x})(\partial V_o(\check{x}',\bar{x};v)/\partial x_i)d\bar{x}d\check{x}' \tag{3.111}$$

$j=1,\ldots,d$, $\check{x} \in D_{o\check{x}}$

$$\Lambda_{1,j}(x;v) \triangleq \int_{D_o} \theta_a^{(m)}(x-x')Q_1(x') \sum_{i=1}^{m} F_{ij}(x')(\partial V_1(x';v)/\partial x_i)dx' \tag{3.112}$$

$j=1,\ldots,d$, $x \in D_o$.

Then

$$\ell(v^o) \le \ell(v) \quad \text{for any } v = (u,\check{v}) \in U_o. \tag{3.113}$$

The determination of v^o by means of (3.101) (or of $(u^o,\check{v}^o)$ by means of (3.110) is in itself a very difficult optimization problem, and, fur= thermore since the establishment of conditions for the existence of solu= tions $\{(V_o(\cdot;v^o),V_1(\cdot;v^o)),\ (Q_o,Q_1),v^o\}$ to the complicated eqns (3.98)-(3.102) (or (3.107)-(3.112)) seems to be even more difficult and there is a lack of any background in the theory of partial differential equa= tions, upon which to build, these problems are not considered here.

A procedure for computing weak suboptimal strategies is suggested in the next section.

3.5 COMPUTATION OF WEAK SUBOPTIMAL STRATEGIES

In this work the following algorithms have been applied to eqns (3.98)-(3.102) or to (3.107)-(3.110) resepctively, in order to compute weak suboptimal strategies.

(a) Eqns (3.98)-(3.102)

1. Given $v^{(o)},v^{(1)},\dots,v^{(n)} \in U_o$
2. Compute $(V_o(\cdot;v^{(n)}),V_1(\cdot;v^{(n)}))$ by solving numerically the following problem:

$$\mathcal{L}_i(v_a^{(n)})(V_o(\alpha),\ V_1(\alpha)) = 0 \quad , \alpha \in D \quad , \quad i=0,1$$

$$(V_o(\alpha;v^{(n)}),V_1(\alpha;v^{(n)}))=(1,1),\alpha\in K;(V_o(\alpha;v^{(n)}),V_1(\alpha;v^{(n)}))=(0,0),\alpha\notin D_o \tag{3.114}$$

3. Calculate $\ell(v^{(n)})$.
4. Compute $(Q_o(\cdot;v^{(n)}),Q_1(\cdot;v^{(n)}))$ by solving numerically the following problem:

$$\begin{aligned} &\mathcal{L}_o^*Q_o(\alpha) + qQ_1(\alpha) = 1 - V_o(\alpha;v^{(n)}),\ \alpha \in D_o \\ &\mathcal{L}_1^*Q_1(\alpha) + qQ_o(\alpha) = 1 - V_1(\alpha;v^{(n)}),\ \alpha \in D_o \\ &(Q_o(\alpha),\ Q_1(\alpha)) = (0,0) \quad , \quad \alpha \notin D_o \end{aligned} \tag{3.115}$$

5. $v^{(n+1)}$ is determined from

$$v_j^{(n+1)}(y) = -v_{o,j}\operatorname{sign}\{\int_{D_{oy}} \theta_a^{(m)}(y-y')\int_{D_{ox}} \sum_{\ell=0}^{1} Q_\ell(x,y';v^{(n)}) \sum_{i=1}^{m} F_{ij}(x)(\partial V_\ell(x,y';v^{(n)})/\partial x_i) \tag{3.116}$$

$dx\ dy'\}$, $j=1,\dots,d$, $y \in D_{oy}$

6. If $v^{(n+1)} \neq v^{(n)}$; then $n + 1 \to n$ and go to 2. Otherwise: stop.

The computations are continued until for some $n \geq 0$ either $v^{(n+1)} = v^{(n)}$ or $\ell(v^{(n+1)}) = \ell(v^{(n)})$.

(b) Eqns (3.107)-(3.110)

In this case, stages 1-4 are similar to stages 1-4 in the case above.

5. $v^{(n+1)} = (u^{(n+1)},\check{v}^{(n+1)})$ is determined from

$$u_j^{(n+1)}(x) = -u_{o,j}\operatorname{sign}\{\int_{D_o} \theta_a^{(m)}(x-x')Q_1(x';v^{(n)}) \sum_{i=1}^{m} F_{ij}(x')(\partial V_1(x';v^{(n)})/\partial x_i)dx'\},$$

$$j=1,\dots,d,\ x \in D_o \tag{3.117}$$

$$\check{v}_j^{(n+1)}(\check{x}) = -\check{v}_{o,j}\operatorname{sign}\{\int_{D_{o\check{x}}} \theta_a^{(p)}(\check{x}-\check{x}')\int_{D_{o\bar{x}}} Q_o(\check{x}',\bar{x};v^{(n)}) \sum_{i=1}^{m} F_{ij}(\check{x},\bar{x})(\partial V_o(\check{x}',\bar{x};v^{(n)})/\partial x_i)d\bar{x}d\check{x}'\}$$

$$j=1,\dots,d,\ x \in D_{ox} \tag{3.118}$$

6. If $(u^{(n+1)},\check{v}^{(n+1)}) \neq (u^{(n)},\check{v}^{(n)})$; then $n + 1 \to n$ and go to 2. Otherwise: stop.

The computations are continued until for some $n \geq 0$ either $(u^{(n+1)},\check{v}^{(n+1)}) = (u^{(n)},\check{v}^{(n)})$ or $\ell((u^{(n+1)},\check{v}^{(n+1)})) = \ell((u^{(n)},\check{v}^{(n)}))$.

Remark 3.2

If the sequence $v^{(n)}$ (eqns (3.114)-(3.116)) converges, and $\lim_{n\to\infty} v^{(n)} = \bar{v}$, then it converges to a solution of the following problem

$$\mathcal{L}_i(v_a)(V_o(\alpha)\ ,\ V_1(\alpha)) = 0\ \ ,\ \alpha \in D\ \ ,\ \ i=0,1 \tag{3.118}$$

$$\mathcal{L}_o^* Q_o(\alpha) + qQ_1(\alpha) = 1 - V_o(\alpha)\ \ ,\ \text{a.e. in } D_o \tag{3.119}$$

$$\mathcal{L}_1^* Q_1(\alpha) + qQ_o(\alpha) = 1 - V_1(\alpha)\ \ ,\ \text{a.e. in } D_o \tag{3.120}$$

$$(V_o(\alpha),V_1(\alpha)) = (1,1),\ \alpha \in K;\ \ (V_o(\alpha),V_1(\alpha)) = (Q_o(\alpha),Q_1(\alpha)) = (0,0),\ \alpha \notin D_o \tag{3.121}$$

$$v_j(y) = -v_{o,j}\text{sign}\{\int_{D_{oy}} \theta_a^{(m)}(y-y')\int_{D_{ox}} \sum_{\ell=0}^{1} Q_\ell(x,y') \sum_{i=1}^{m} F_{ij}(x)(\partial V_\ell(x,y')/\partial x_i)dxdy'\}, \tag{3.122}$$

$$j=1,\dots,d, \quad y \in D_{oy}.$$

For all the examples which have been numerically solved here, the results indicate that

$$\ell(v^{(o)}) \geq \ell(v^{(1)}) \geq \dots \geq \ell(v^{(n)}) \geq \dots$$

Unfortunately, owing to the complexity of eqns (3.98)-(3.102), and of the algorithm for computing weak suboptimal strategies, we cannot here give condi= tions for the existence of $\lim_{n\to\infty} v^{(n)} = \bar{v}$, and whenever $\{v^{(n)}\}$ converges to $\bar{v}$ we cannot determine whether $\bar{v}$ is also a weak optimal strategy. The same ar= gument can be applied to the sequence $\{(u^{(n)},\check{v}^{(n)})\}$, obtained via the implemen= tation of the algorithm for computing weak suboptimal strategies on eqns (3.107)-(3.110).

3.6 THE NUMERICAL METHOD

In this section, a finite-differences scheme is described for solving numerically (3.114) and (3.115) (for a given strategy $v \in U_o$), where $\mathcal{L}_i(v_a)$, i=0,1, and $\mathcal{L}_i^*$, i=0,1, are given by (3.19)-(3.20) and (3.86)-(3.87) ($\mathcal{L}_i^*$ is the adjoint operator of $\mathcal{L}_i$, i=0,1) respectively. It is assumed that f, F, σ and γ satisfy all the conditions stated in Section 3.4. It is fur= ther assumed that $\bar{\sigma}_{ij}(x) = (\sigma(x)\sigma'(x))_{ij} = \delta_{ij}\,\bar{\sigma}_{ii}$ and $\bar{\gamma}_{ij}(x) = (\gamma(x)\gamma'(x))_{ij} = \delta_{ij}\,\bar{\gamma}_{ii}$, $x \in \mathbb{R}^m$, i,j=1,...,m, which is the case for all the examples here solved numerically.

Let $\mathbb{R}_h^{2m}$ be a finite-difference grid on $\mathbb{R}^{2m}$, with a constant mesh size h along all axes. Denote $\alpha = (x,y) \in \mathbb{R}^m \times \mathbb{R}^m$ and define $D_h \triangleq \mathbb{R}_h^{2m} \cap D$, $D_{oh} \triangleq \mathbb{R}_h^{2m} \cap D_o$ and $K_h \triangleq \mathbb{R}_h^{2m} \cap K$. Also, denote by e^i the unit vector along the i-th axis, i=1,...,2m. By applying a procedure similar to that described by (2.137)-(2.141), eqns (3.114) and (3.115) yield

$$V_i(\alpha) = (F_i(V_o,V_1))(\alpha) \quad , \quad i=0,1, \alpha \in D_h \tag{3.123}$$

$$(V_o(\alpha),V_1(\alpha)) = (1,1),\ \alpha \in K_h\ ;\ (V_o(\alpha),V_1(\alpha)) = (0,0),\ \alpha \notin D_{oh} \quad (3.124)$$

$$S_i(\alpha)=(S_i(Q_o,Q_1,V_i))(\alpha),\ i=0,1,\alpha \in D_{oh};\ (Q_o(\alpha),Q_1(\alpha))=(0,0),\ \alpha \notin D_{oh} \quad (3.125)$$

where

$$(F_o(V_o,V_1))(\alpha) \triangleq \sum_{i=1}^{m} \{P_{o,i}(\alpha;v)V_o(\alpha+he^i) + P_{o,-i}(\alpha;v)V_o(\alpha-he^i)$$

$$+ P_{o,i+m}(\alpha;v)V_o(\alpha+he^{i+m})+P_{o,-(i+m)}(\alpha;v)V_o(\alpha-he^{i+m})\},$$

$$+ qh^2V_1(\alpha)/R_o(\alpha;v) \quad , \qquad (3.126)$$

$$(F_1(V_o,V_1))(\alpha) \triangleq \sum_{i=1}^{m} \{P_{1,i}(\alpha;v)V_1(\alpha+he^i) + P_{1,-i}(\alpha;v)V_1(\alpha-he^i)$$

$$+ P_{1,i+m}(\alpha;v)V_1(\alpha+he^{i+m})+P_{1,-(i+m)}(\alpha;v)V_1(\alpha-he^{i+m})\}$$

$$+ qh^2V_o(\alpha)/R_1(\alpha;v), \qquad (3.127)$$

$$(S_o(Q_o,Q_1,V_o))(\alpha) \triangleq \sum_{i=1}^{m} \{S_{o,i}(x)Q_o(\alpha+he^i)+S_{o,-i}(x)Q_o(\alpha-he^i)$$

$$+ S_{o,i+m}(x)Q_o(\alpha+he^{i+m})+S_{o,-(i+m)}(x)Q_o(\alpha-he^{i+m})\}$$

$$+ qh^2Q_1(\alpha)/S_o(x) - h^2(1 - V_o(\alpha))/S_o(x), \qquad (3.128)$$

$$(S_1(Q_o,Q_1,V_1))(\alpha) \triangleq \sum_{i=1}^{m} \{S_{1,i}(x)Q_1(\alpha+he^i) + S_{1,-i}(x)Q_1(\alpha-he^i)$$

$$+ S_{1,i+m}(x)Q_1(\alpha+he^{i+m}) + S_{1,-(i+m)}(x)Q_1(\alpha-he^{i+m})\}$$

$$+ qh^2Q_o(\alpha)/S_1(x) - h^2(1 - V_1(\alpha))/S_1(x); \qquad (3.129)$$

and

$$R_o(\alpha;v) \triangleq \sum_{i=1}^{m} [\bar{\sigma}_{ii} + \bar{\gamma}_{ii} + h \mid F_i(\alpha;v)|] + qh^2 \qquad (3.130)$$

$$F_i(\alpha;v) \triangleq f_i(x) + \sum_{j=1}^{d} F_{ij}(x)v_j(y), \quad i=1,\dots,m \tag{3.131}$$

$$P_{o,i}(\alpha;v) \triangleq [\bar{\sigma}_{ii}/2 + h \max(F_i(\alpha;v),0)]/R_o(\alpha;v) \tag{3.132}$$

$$P_{o,-i}(\alpha;v) \triangleq [\bar{\sigma}_{ii}/2 - h \min(F_i(\alpha;v),0)]/R_o(\alpha;v) \tag{3.133}$$

$$P_{o,i+m}(\alpha;v) \triangleq P_{o,-(i+m)}(\alpha;v) \triangleq \bar{\gamma}_{ii}/(2R_o(\alpha;v)) \tag{3.134}$$

$$R_1(\alpha;v) \triangleq \sum_{i=1}^{m} [\bar{\sigma}_{ii} + \bar{\gamma}_{ii} + h(|F_i(\alpha;v)| + |x_i|)] + qh^2 \tag{3.135}$$

$$P_{1,i}(\alpha;v) \triangleq [\bar{\sigma}_{ii}/2 + h \max(F_i(\alpha;v),0)]/R_1(\alpha;v) \tag{3.136}$$

$$P_{1,-i}(\alpha;v) \triangleq [\bar{\sigma}_{ii}/2 - h \min(F_i(\alpha;v),0)]/R_1(\alpha;v) \tag{3.137}$$

$$P_{1,i+m}(\alpha;v) \triangleq [\bar{\gamma}_{ii}/2 + h \max(x_i,0)]/R_1(\alpha;v) \tag{3.138}$$

$$P_{1,-(i+m)}(\alpha;v) \triangleq [\bar{\gamma}_{ii}/2 - h \min(x_i,0)]/R_1(\alpha;v) \tag{3.139}$$

$$S_o(x) \triangleq \sum_{i=1}^{m} [\bar{\sigma}_{ii} + \bar{\gamma}_{ii} + h(|f_i(x)| + (\tfrac{1}{2})(f_i(x+he^i)-f_i(x-he^i)))] + qh^2 \tag{3.140}$$

$$S_{o,i}(x) \triangleq [\bar{\sigma}_{ii}/2 - h \min(f_i(x),0)]/S_o(x) \tag{3.141}$$

$$S_{o,-i}(x) \triangleq [\bar{\sigma}_{ii}/2 + h \max(f_i(x),0)]/S_o(x) \tag{3.142}$$

$$S_{o,i+m}(x) \triangleq S_{o,-(i+m)}(x) \triangleq \bar{\gamma}_{ii}/(2S_o(x)) \tag{3.143}$$

$$S_1(x) \triangleq \sum_{i=1}^{m} [\bar{\sigma}_{ii} + \bar{\gamma}_{ii} + h(|f_i(x)| + |x_i| + (\tfrac{1}{2})(f_i(x+he^i)-f_i(x-he^i)))] + qh^2 \tag{3.144}$$

$$S_{1,i}(x) \triangleq [\bar{\sigma}_{ii}/2 - h \min(f_i(x),0)]/S_1(x) \tag{3.145}$$

$$S_{1,-i}(x) \triangleq [\bar{\sigma}_{ii}/2 + h \max(f_i(x),0)]/S_1(x) \tag{3.146}$$

$$S_{1,i+m}(x) \triangleq [\bar{\gamma}_{ii}/2 - h \min(x_i,0)]/S_1(x) \tag{3.147}$$

$$S_{1,-(i+m)}(x) \triangleq [\bar{\gamma}_{ii}/2 + h \max(x_i,0)]/S_1(x). \tag{3.148}$$

Eqns (3.123)-(3.124), and eqns (3.125) (for a given (V_o,V_1)) are solved by an iterative procedure using the underrelaxation technique with an acceleration factor w_o, until the difference between two consecutive iterations does not exceed a given tolerance ε_o.

Given $h > 0$ we choose $0 < a << h$, (see (3.73)). Then, whenever the algorithm for computing weak suboptimal strategies, where the scheme given by (3.123)-(3.148) is applied, converges to a unique solution we denote it by $\{V_o^h(\cdot;\bar{v}^h),V_1^h(\cdot;\bar{v}^h),\ Q_o^h(\cdot;\bar{v}^h),\ Q_1^h(\cdot;\bar{v}^h),\ \bar{v}^h\}$.

A finite-difference procedure, similar to that given here, can be derived for the case where the system is given by eqns (3.1) and (3.11). In this case, whenever the algorithm for computing weak suboptimal stra= tegies converges, we denote the limit solution by $\{V_o^h(\cdot;\bar{v}^h),V_1^h(\cdot;\bar{v}^h),$ $Q_o^h(\cdot;\bar{v}^h),Q_1^h(\cdot;\bar{v}^h),\ \bar{v}^h = (\bar{u}^h,\check{\bar{v}}^h)\}$.

3.7 EXAMPLE 1 : SMOOTH STRATEGIES FOR STEERING A RANDOM MOTION OF A POINT ALONG A LINE

The following example is taken from Yavin and Venter [38].

3.7.1 Statement of the problem

Consider the dynamical system given by

$$dx = v(y)dt + \sigma dW\ ,\ t > 0,\ x \in \mathbb{R} \tag{3.149}$$

with the observation

$$dy = \theta x dt + \gamma dB,\quad t > 0,\quad y \in \mathbb{R} \tag{3.150}$$

where $W = \{W(t),\ t \geq 0\}$ and $B = \{B(t),\ t \geq 0\}$ are $\mathbb{R}$-valued standard Wiener processes. $\Theta = \{\theta(t),\ t \geq 0\}$ is a homogeneous jump Markov process with state space $S = \{0,1\}$ as described in Section 3.1. We assume that W,B and Θ are mutually independent. Define

$$D_o \overset{\Delta}{=} \{\alpha = (x,y) : |x| < 1 \text{ and } |y| < 1\} \tag{3.151}$$

$$K \overset{\Delta}{=} \{\alpha = (x,y) : |x| \le \rho \text{ and } |y| \le 1 - \delta\} \tag{3.152}$$

$$D \overset{\Delta}{=} D_o - K \tag{3.153}$$

where $0 < \rho < 1$ and $0 < \delta << 1$ are given numbers. Denote by $\zeta_\alpha^v = \{\zeta_\alpha^v(t) = (\zeta_{\alpha,1}^v(t), \eta_{\alpha,1}^v(t)),\ t \ge 0\}$ the solution to (3.149)-(3.150) where $\alpha = (x,y) \in \mathbb{R}^2$, $v \in U$ and let $\tau_i(\alpha;v)$, $i=0,1$ and U be defined as in (3.4) and (3.7) respectively. Denote

$$V_i(\alpha;v) \overset{\Delta}{=} P_{\alpha,i}(\{\zeta_\alpha^v(\tau_i(\alpha;v)) \in K\}) - E_{\alpha,i} \int_0^{\tau_i(\alpha;v)} \lambda(\theta(t))v^2(\eta_{\alpha,1}^v(t))dt \tag{3.154}$$

$v \in U$, $i=0,1$.

where $\lambda : \{0,1\} \to \mathbb{R}$ is a given function satisfying $0 < \lambda(o) \le \lambda(1)$. $P_{\alpha,i}(\{\zeta_\alpha^v(\tau_i(\alpha;v)) \in K\})$ is the probability that $\zeta_\alpha^v(t)$ reaches the set K before it leaves D_o.

Note that eq (3.154) is a special case of (3.8). The problem considered in this example is to maximize the functional J on U where

$$J(v) = \int_{D_o} \sum_{i=o}^{1} P(\theta(o)=i)V_i(\alpha;v)d\alpha = \int_{D_o} \sum_{i=o}^{1} \pi_i\ V_i(\alpha;v)d\alpha. \tag{3.155}$$

Assuming that there exists a strategy $v^o \in U$ such that $J(v^o) \ge J(v)$ for any $v \in U$, and that all the conditions stated in Theorem 3.1 are satisfied, it follows from equations (3.41),(3.42),(3.44) and (3.149)-(3.150) that the following set of equations has to be solved:

$$v(y)\partial V_o(\alpha)/\partial x + (\tfrac{1}{2})(\sigma^2\partial^2 V_o(\alpha)/\partial x^2 + \gamma^2\partial^2 V_o(\alpha)/\partial y^2) - qV_o(\alpha) + qV_1(\alpha) = \lambda(o)v^2(y) \quad , \alpha \in D, \tag{3.156}$$

$$v(y)\partial V_1(\alpha)/\partial x + x\partial V_1(\alpha)/\partial y + (\tfrac{1}{2})(\sigma^2\partial^2 V_1(\alpha)/\partial x^2 + \gamma^2\partial^2 V_1(\alpha)/\partial y^2)$$

$$-qV_1(\alpha) + qV_0(\alpha) = \lambda(1)v^2(y) \quad , \alpha \in D, \tag{3.157}$$

$$-v(y)\partial Q_0(\alpha)/\partial x + (\tfrac{1}{2})(\sigma^2\partial^2 Q_0(\alpha)/\partial x^2 + \gamma^2\partial^2 Q_0(\alpha)/\partial y^2) - qQ_0(x) = -1 \tag{3.158}$$

$\alpha \in D_0$,

$$-v(y)\partial Q_1(\alpha)/\partial x - x\partial Q_1(\alpha)/\partial y + (\tfrac{1}{2})(\sigma^2\partial^2 Q_1(\alpha)/\partial x^2 + \gamma^2\partial^2 Q_1(\alpha)/\partial y^2)$$

$$-qQ_1(\alpha) = -1 \quad , \alpha \in D_0, \tag{3.159}$$

$$V_0(\alpha) = V_1(\alpha) = 1,\ \alpha \in K; \quad V_0(\alpha) = V_1(\alpha) = Q_0(\alpha) = Q_1(\alpha) = 0,\ \alpha \notin D_0, \tag{3.160}$$

$$v(y) = (\tfrac{1}{2})\int_{-1}^{1} \{\pi_0 Q_0(\alpha)\partial V_0(\alpha)/\partial x + \pi_1 Q_1(\alpha)\partial V_1(\alpha)/\partial x\}dx \Big/ \int_{-1}^{1} [\lambda(o)\pi_0 Q_0(\alpha)$$

$$+ \lambda(1)\pi_1 Q_1(\alpha)]I_D(\alpha)dx. \tag{3.161}$$

3.7.2 Results

Eqns (3.156)-(3.161) were solved, using a finite-difference scheme similar to that described in Section 3.6. Given $h > 0$, we denote by $(V_0^h, V_1^h, Q_0^h, Q_1^h, \bar{v}^h)$ the solution, whenever it exists, to the set of finite differences equations.

Using the finite-difference scheme, eqns (3.156)-(3.161) were solved and the values of $(V_0^h, V_1^h, Q_0^h, Q_1^h, \bar{v}^h)$ computed for the following set of parameters: $\delta = 10^{-4}$, $\rho = 0.1$, $\sigma^2 = 5\cdot 10^{-4}$, $5\cdot 10^{-10} \le \gamma^2 \le 0.5$, $\pi_0=\pi_1=0.5$, $0.5 \le \lambda(o) \le \lambda(1) \le 8$, $0.5 \le q \le 16$, $\varepsilon_0 = 10^{-4}$ (ε_0 is the tolerance be= tween two consecutive iterations, see at the end of Section 3.6), and $h=0.05$. In addition, eqns (3.156)-(3.157) (together with the boundary conditions on V_i, $i=0,1$, (3.160)) were solved, using the finite-difference scheme, for $\lambda(o) = \lambda(1) = 0$ and $v = \bar{v}^h$ yielding a solution (P_o^h, P_1^h) with the following probabilistic interpretation

$P_i^h(\alpha)$ is the approximation on D_{oh} to $P_{\alpha,i}(\{\zeta_\alpha^{\bar{v}}(\tau_i(\alpha;\bar{v})) \in K\})$, i=0,1

where $\bar{v} \in U$ is obtained by interpolating $\bar{v}^h$.

In order to assess the accuracy of the numerical method, the values of $(V_o^h, V_1^h, Q_o^h, Q_1^h, \bar{v}^h, P_o^h, P_1^h)$ were computed for h=0.1, 0.05, 0.025; the results are given in Tables 3.1 and 3.2.

TABLE 3.1: The values of $\int_{D_o} V_i^h(\alpha)d\alpha$, $\int_{D_o} Q_i^h(\alpha)d\alpha$, $\int_{D_o} P_i^h(\alpha)d\alpha$, i=0,1, for various values of h, where : $\gamma^2 = 5\cdot 10^{-4}$, $\lambda(o) = \lambda(1) = 1$ and q=1.

h	$\int_{D_o} V_o^h(\alpha)d\alpha$	$\int_{D_o} V_1^h(\alpha)d\alpha$	$\int_{D_o} Q_o^h(\alpha)d\alpha$	$\int_{D_o} Q_1^h(\alpha)d\alpha$	$\int_{D_o} P_o^h(\alpha)d\alpha$	$\int_{D_o} P_1^h(\alpha)d\alpha$
0.1	.3164	.3084	.8806	.6997	.3238	.3144
0.05	.3143	.3080	.9258	.7352	.3224	.3147
0.025	.3133	.3083	.9463	.7527	.3213	.3149

TABLE 3.2: The values of $\int_{D_o} V_i^h(\alpha)d\alpha$, $\int_{D_o} Q_i^h(\alpha)d\alpha$, $\int_{D_o} P_i^h(\alpha)d\alpha$, i=0,1, for various values of h, where $\gamma^2 = 5\cdot 10^{-4}$, $\lambda(o) = \lambda(1) = 8$ and q=1.

h	$\int_{D_o} V_o^h(\alpha)d\alpha$	$\int_{D_o} V_1^h(\alpha)d\alpha$	$\int_{D_o} Q_o^h(\alpha)d\alpha$	$\int_{D_o} Q_1^h(\alpha)d\alpha$	$\int_{D_o} P_o^h(\alpha)d\alpha$	$\int_{D_o} P_1^h(\alpha)d\alpha$
0.1	.1879	.1854	.8959	.6931	.1892	.1864
0.05	.1780	.1757	.9401	.7277	.1792	.1767
0.025	.1734	.1713	.9589	.7446	.1747	.1724

The sensitivity of $(V_0^h, V_1^h, Q_0^h, Q_1^h, P_0^h, P_1^h)$ to variations in the value of γ^2(the noise factor in the observation) is demonstrated in Table 3.3.

TABLE 3.3: The values of $\int_{D_0} V_i^h(\alpha)d\alpha$, $\int_{D_0} Q_i^h(\alpha)d\alpha$, $\int_{D_0} P_i^h(\alpha)d\alpha$, i=0,1, for various values of γ^2, where : $\lambda(o) = \lambda(1) = 1$, q=1 and h=0.05.

γ^2	$\int_{D_0} V_0^h(\alpha)d\alpha$	$\int_{D_0} V_1^h(\alpha)d\alpha$	$\int_{D_0} Q_0^h(\alpha)d\alpha$	$\int_{D_0} Q_1^h(\alpha)d\alpha$	$\int_{D_0} P_0^h(\alpha)d\alpha$	$\int_{D_0} P_1^h(\alpha)d\alpha$
$5\cdot10^{-10}$	.3156	.3091	.9299	.7359	.3237	.3157
$5\cdot10^{-8}$	.3156	.3091	.9299	.7359	.3237	.3157
$5\cdot10^{-6}$	.3156	.3091	.9298	.7359	.3237	.3157
$5\cdot10^{-4}$	.3143	.3080	.9258	.7352	.3224	.3147
$5\cdot10^{-2}$	.2298	.2289	.8062	.6823	.2310	.2299
$5\cdot10^{-1}$	.1387	.1385	.5009	.4643	.1387	.1385

In Fig.3.1 four samples of the control law $\bar{v}^h = \bar{v}^h(y)$ are given. These are the cases where: (i) $\gamma^2 = 5\cdot10^{-4}$, $\lambda(o) = 1$, $\lambda(1) = 0.1$, q=1 and h=0.05. (ii) $\gamma^2 = 5\cdot10^{-4}$, $\lambda(o) = \lambda(1) = 0.8$, q=1 and h=0.05. (iii) $\gamma^2 = 5\cdot10^{-4}$, $\lambda(o) = \lambda(1) = 1$, q=1 and h=0.05. (iv)$\gamma^2 = 5\cdot10^{-4}$, $\lambda(o) = \lambda(1) = 1$,q=8 and h=0.05.

Figs. 3.2A and 3.2B show $\int_{D_0} V_i^h(\alpha)d\alpha$, $\int_{D_0} P_i^h(\alpha)d\alpha$, i=0,1, as functions of $\lambda = \lambda(o) = \lambda(1)$ for the cases where: $\gamma^2 = 5\cdot10^{-4}$, q=1 and h=0.05. The plots show that $\int_{D_0} V_i^h(\alpha)d\alpha$ and $\int_{D_0} P_i^h(\alpha)d\alpha$, i=0,1, decrease when λ increases. Fig.3.3 shows $\int_{D_0} P^h(\alpha)d\alpha$ as a function of q for the case where: $\gamma^2 = 5\cdot10^{-4}$, $\lambda(o) = \lambda(1) = 1$ and h=0.05; where $P^h = \pi_0 P_0^h + \pi_1 P_1^h$. The plot shows that $\int_{D_0} P^h(\alpha)d\alpha$ decreases when q increases.

Fig.3.1 shows the typical form of the control law $\bar{v}^h$, and Figs.3.2 and 3.3 show the trend of all the results obtained here, which is that V_i^h and P_i^h,i=0,1, decrease when $\lambda = \lambda(o) = \lambda(1)$ or q increase.

Finally, the values of $\int_{D_0} Q_i^h(\alpha)d\alpha$,i=0,1, as functions of q, in the case where $\gamma^2 = 5\cdot10^{-4}$, $\lambda(o) = \lambda(1) = 1$ and h=0.05, have been plotted in Fig.3.4. Again, these values decrease when q increases.

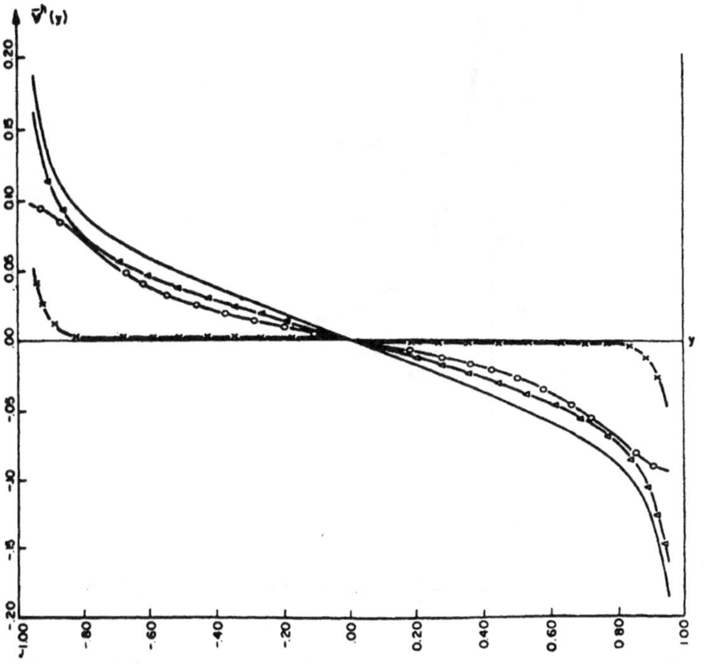

Fig.3.1: The control law $\bar{v}^h = \bar{v}^h(y)$ for four cases. (i) ———— $\sigma_2^2 = 5\cdot 10^{-4}$, $\lambda(o) = 1$, $\lambda(1) = 0.1$, q=1 and h=0.05. (ii) - Δ - Δ - Δ - Δ $\sigma_2^2 = 5\cdot 10^{-4}$, $\lambda(o) = \lambda(1) = 0.8$, q=1 and h=0.05. (iii) —o—o—o—o $\sigma_2^2 = 5\cdot 10^{-4}$, $\lambda(o) = \lambda(1) = 1$, q=1 and h=0.05. (iv) —x—x—x—x $\sigma_2^2 = 5\cdot 10^{-4}$, $\lambda(o) = \lambda(1) = 1$, q=8 and h=0.05.

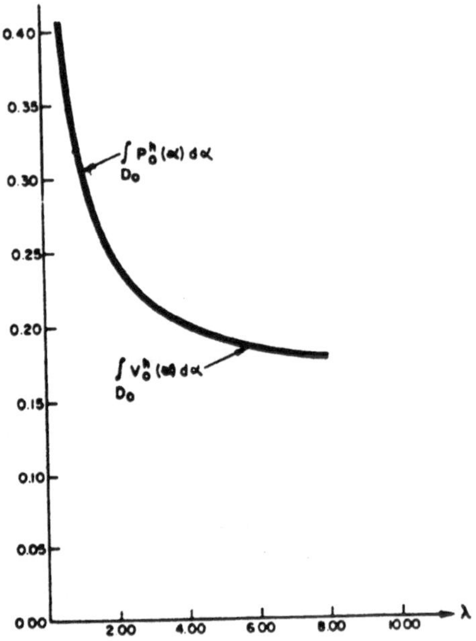

Fig.3.2-A: $\int_{D_o} V_o^h(\alpha)d\alpha$ and $\int_{D_o} P_o^h(\alpha)d\alpha$ as functions of $\lambda = \lambda(o) = \lambda(1)$, where $\gamma^2 = 5 \cdot 10^{-4}$, q=1 and h=0.05.

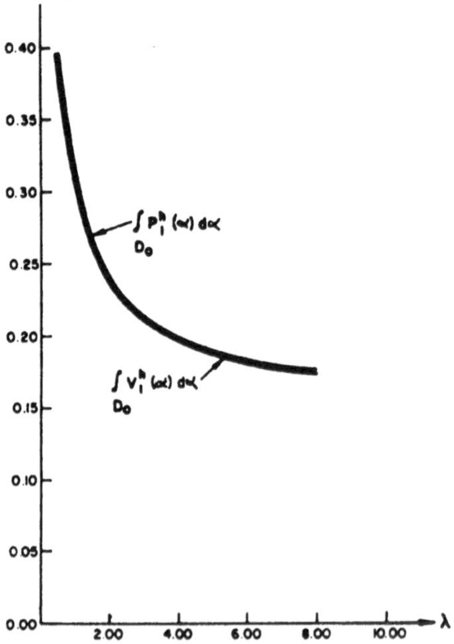

Fig.3.2-B: $\int_{D_o} V_1^h(\alpha)d\alpha$ and $\int_{D_o} P_1^h(\alpha)d\alpha$ as functions of $\lambda = \lambda(o) = \lambda(1)$, where $\gamma^2 = 5 \cdot 10^{-4}$, q=1 and h=0.05.

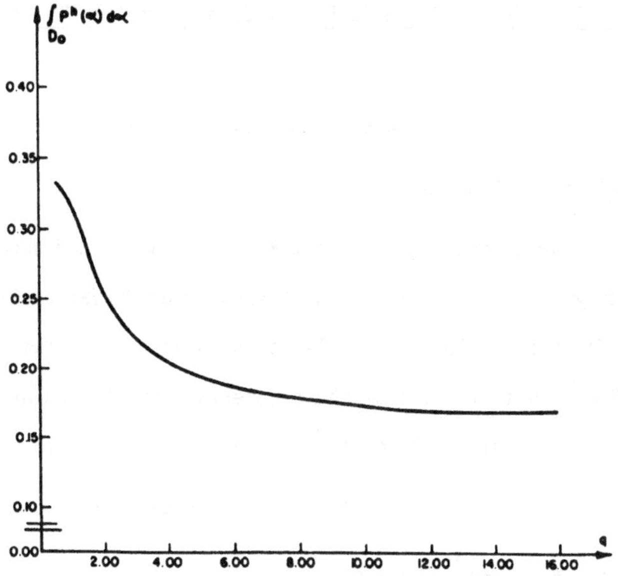

Fig.3.3: $\int_{D_0} P^h(\alpha)d\alpha$ as a function of q, $P^h = \pi_0 P_0^h + \pi_1 P_1^h$, where $\gamma^2 = 5\cdot 10^{-4}$, $\lambda(o) = \lambda(1) = 1$ and h=0.05.

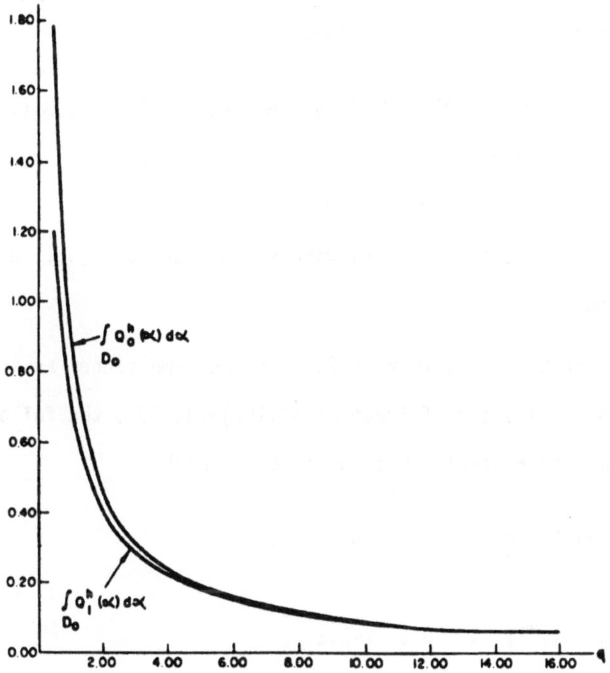

Fig.3.4: $\int_{D_0} Q_i^h(\alpha)d\alpha$, i=0,1, as functions of q, where $\gamma^2 = 5\cdot 10^{-4}$, $\lambda(o) = \lambda(1) = 1$ and h=0.05.

3.8 EXAMPLE 2: BANG-BANG STRATEGIES FOR STEERING A RANDOM MOTION OF A POINT

The following example is taken from Yavin [39].

3.8.1 Statement of the Problem

Consider a random motion of a point M in the plane, and suppose that the velocity $v = (v_1, v_2)$ of M is perturbed by an $\mathbb{R}^2$-valued Gaussian white noise. The observations available to the point are of the kind as given by (3.2). With the aid of these observations, the point M wishes to steer itself into a given closed neighbourhood of the origin.

Let (Ω, F, P) be a probability space. The motion of M is given by

$$dx_i = v_i(y)dt + \sigma_i dW_i \quad , t > 0 \ , i=1,2 \quad , \quad y = (y_1, y_2) \tag{3.162}$$

and the interrupted noisy observations are given by

$$dy_i = \theta x_i dt + \gamma_i dB_i \quad , t > 0 \quad , i=1,2 \tag{3.163}$$

where $W = \{W(t) = (W_1(t), W_2(t)),\ t \geq 0\}$ and $B = \{B(t) = (B_1(t), B_2(t)),\ t \geq 0\}$ are $\mathbb{R}^2$-valued standard Wiener processes on (Ω, F, P), and σ_i, $i=1,2$, and γ_i, $i=1,2$, are given positive numbers. $\Theta = \{\theta(t),\ t \geq 0\}$ is a homo= geneous jump Markov process on (Ω, F, P) with state space $S = \{0,1\}$ as de= scribed in Section 3.1.

Let U be given by (2.37) (here d=2). In the same manner as in Sections 3.3 or 3.4 , instead of treating (3.162)-(3.163), the following set of stochastic differential equations is dealt with:

$$\begin{cases} dx_i = v_{a,i}(y)dt + \sigma_i dW_i \ , \quad t > 0 \ , \ i=1,2 \\ dy_i = \theta x_i dt + \gamma_i dB_i, \ t > 0 \ , \quad i=1,2 \end{cases} \tag{3.164}$$

where $v_a = (v_{a,1}, v_{a,2})'$ is given by (3.73) (here m=2),and $a \ll 1$. Define the following sets in $\mathbb{R}^4$:

$$D_o \triangleq \{\alpha = (x,y) : |x_i| < 1 \text{ and } |y_i| < 1,\ i=1,2\} \tag{3.165}$$

$$K \triangleq \{\alpha = (x,y) : x_1^2 + x_2^2 \le \rho^2 ,\ |y_i| \le 1 - \delta,\ i=1,2\} \tag{3.166}$$

$$D \triangleq D_o - K \tag{3.167}$$

where $0 < \rho < 1$ and $0 < \delta << 1$ are given numbers. Denote by ζ_α^v, $\zeta_\alpha^v = \{\zeta_\alpha^v(t) = (\zeta_{\alpha,1}^v(t), \zeta_{\alpha,2}^v(t), \eta_{\alpha\ ,1}^v(t), \eta_{\alpha,2}^v(t)),\ t \ge 0\}$, $v \in \tilde{U}$ (see Section 3.4 for the definition of $\tilde{U}$, $\tau_i(\cdot;v)$, $i=0,1$, U and U_o) the solu= tion to (3.164). Also, let

$$V_i(\alpha;v) \triangleq P_{\alpha,i}(\{\zeta_\alpha^v(\tau_i(\alpha;v)) \in K\}),\ i=0,1,\ v \in U, \alpha \in \bar{D}_o \quad , \tag{3.76}$$

and

$$\ell(v) \triangleq \sum_{i=o}^{1} \int_{D_o} (1 - V_i(\alpha;v))^2 d\alpha,\ v \in U_o \tag{3.77}$$

In this section the following problem is considered: find a strategy $v^o \in U_o$ such that $\ell(v^o) \le \ell(v)$ for any $v \in U_o$.

A strategy $v^o \in U_o$, whenever it exists, is supposed, roughly speak= ing, to steer the state $\zeta_\alpha^{v^o}(t)$ in such a manner as to maximize, in the $L_2(D_o)$ sense, the probability that the state reaches the set K before reaching ∂D_o.

In this example the algorithm for computing weak suboptimal strate= gies takes the following form:

1. Given $v^{(n)} \in U_o$
2. Calculate $v_a^{(n)}$ (eq.(3.73)).
3. Compute $(V_o(\cdot;v^{(n)}), V_1(\cdot;v^{(n)}))$ by solving numerically the following problem:

$$\sum_{i=1}^{2} v_{a,i}^{(n)}(y)\partial V_o(\alpha)/\partial x_i + (\tfrac{1}{2}) \sum_{i=1}^{2} (\sigma_i^2\ \partial^2 V_o(\alpha)/\partial x_i^2 + \gamma_i^2\ \partial^2 V_o(\alpha)/\partial y_i^2)$$

(3.168)

$$-qV_o(\alpha) + qV_1(\alpha) = 0 \quad , \alpha \in D$$

$$\sum_{i=1}^{2} (v_{a,i}^{(n)}(y)\partial V_1(\alpha)/\partial x_i + x_i \partial V_1(\alpha)/\partial y_i)$$

(169)

$$+ (\tfrac{1}{2}) \sum_{i=1}^{2} (\sigma_i^2\ \partial^2 V_1(\alpha)/\partial x_i^2 + \gamma_i^2\ \partial^2 V_1(\alpha)/\partial y_i^2) - qV_1(\alpha) + qV_o(\alpha)=0, \ \alpha \in D$$

$$V_o(\alpha) = V_1(\alpha) = 1,\ \alpha \in K\ ;\quad V_o(\alpha) = V_1(\alpha) = 0,\ \alpha \notin D_o. \qquad (3.170)$$

4. Calculate $\ell(v^{(n)})$

5. Compute $(Q_o(\cdot;v^{(n)}),\ Q_1(\cdot;v^{(n)}))$ by solving numerically the following problem:

$$(\tfrac{1}{2}) \sum_{i=1}^{2} (\sigma_i^2\ \partial^2 Q_o(\alpha)/\partial x_i^2 + \gamma_i^2\ \partial^2 Q_o(\alpha)/\partial y_i^2) - qQ_o(\alpha) + qQ_1(\alpha)$$

(3.171)

$$= 1 - V_o(\alpha;v^{(n)}) \quad , \alpha \in D_o$$

$$-\sum_{i=1}^{2} x_i \partial Q_1(\alpha)/\partial y_i + (\tfrac{1}{2}) \sum_{i=1}^{2} (\sigma_i^2\ \partial^2 Q_1(\alpha)/\partial x_i^2 + \gamma_i^2\ \partial^2 Q_1(\alpha)/\partial y_i^2)$$

(3.172)

$$-qQ_1(\alpha) + qQ_o(\alpha) = 1 - V_1(\alpha;v^{(n)}) \quad , \alpha \in D_o$$

$$Q_o(\alpha) = Q_1(\alpha) = 0,\ \alpha \notin D_o \qquad (3.173)$$

6. $v^{(n+1)}$ is determined by

$$v_i^{(n+1)}(y') = -v_{o,i}\ \mathrm{sign}\{\int_{D_{oy}} \theta_a^{(2)}(y-y') \int_{D_{ox}} \sum_{j=o}^{1} Q_j(\alpha;v^{(n)})(\partial V_j(\alpha;v^{(n)})/\partial x_i)dxdy\}$$

(3.174)

$$i=1,2 \quad , \quad y' \in D_{oy}$$

7. If $v^{(n+1)} \neq v^{(n)}$; then $n + 1 \rightarrow n$ and go to 2. Otherwise: stop.

The computations are continued until for some $n \geq 0$ either $v^{(n+1)} = v^{(n)}$ or $\ell(v^{(n+1)}) = \ell(v^{(n)})$.

Note that here: $D_{ox} = \{x : |x_i| < 1,\ i=1,2\}$ and $D_{oy} = \{y : |y_i| < 1,\ i=1,2\}$.

3.8.2 Results

The algorithm for computing weak suboptimal strategies has been applied here to the following set of parameters: $\sigma_1^2 = \sigma_2^2 = 10^{-5}, 10^{-6}$; $\gamma_1^2 = \gamma_2^2 = 10^{-3}, 10^{-4}$; $\varepsilon_o = 10^{-4}$ (see Section 3.6); $\rho = 0.1$; $v_{o,1} = v_{o,2} = v_o = 0.0, 0.1, 0.15, 0.2, 0.3$; $q = 0.5, 1.0, 2,4,6,8,10$; $h=1/6$ and $a << h$. In all the cases computed $v^{(n)}$ converged to $\bar{v}^h$ for $n \leq 10$. In all these cases it also turned out the $\bar{v}^h$ was given by

$$\bar{v}_i^h(y) = -v_o \operatorname{sign}(y_i),\ i=1,2 \tag{3.175}$$

where $\operatorname{sign}(0) = 0$.

It may always be assumed that during the observation process $y(o) = (0,0)$. Consequently, only the values of $V_i^h(x_1,x_2,0,0;\bar{v}^h)$, $i=0,1$, will be presented in this section.

For all the cases computed there it turned out (as expected) that

$$V_i^h(-x_1,x_2,0,0;\bar{v}^h) = V_i^h(x_1,x_2,0,0;\bar{v}^h),\ i=0,1,\ x \in D_{ox} \cap \mathbb{R}_h^2 \tag{3.176}$$

$$V_i^h(x_1,-x_2,0,0;\bar{v}^h) = V_i^h(x_1,x_2,0,0;\bar{v}^h),\ i=0,1,\ x \in D_{ox} \cap \mathbb{R}_h^2. \tag{3.177}$$

Typical extracts from the numerical results are presented in the tables below.

Let $\alpha^{(i,j)} = (ih,jh,0,0)$, $h = 1/6$, $i,j = 0, \pm 1,\dots,\pm 6$. Tables 3.4 - 3.7 show the values of $V_i^h(\alpha^{(\ell,m)};\bar{v}^h)$, $i=0,1$, for $\ell = 0,\dots,5$ and $m=0,2,4$.

TABLE 3.4: $V_0^h(\alpha^{(\ell,m)};\bar{v}^h)$ for $\sigma_1^2 = \sigma_2^2 = 10^{-5}$, $\gamma_1^2 = \gamma_2^2 = 10^{-3}$, q=1, v_0=0.2

	ℓ = 0	ℓ = 1	ℓ = 2	ℓ = 3	ℓ = 4	ℓ = 5
m = 4	.4789	.1983	.1935	.1638	.1200	.0780
m = 2	.8542	.4071	.3789	.2933	.1935	.1133
m = 0	1.0000	.9135	.8542	.6889	.4789	.2963

TABLE 3.5: $V_1^h(\alpha^{(\ell,m)};\bar{v}^h)$ for $\sigma_1^2 = \sigma_2^2 = 10^{-5}$, $\gamma_1^2 = \gamma_2^2 = 10^{-3}$, q=1, v_0=0.2.

	ℓ = 0	ℓ = 1	ℓ = 2	ℓ = 3	ℓ = 4	ℓ = 5
m = 4	.4917	.2002	.1957	.1656	.1213	.0788
m = 2	.8782	.4138	.3851	.2975	.1957	.1141
m = 0	1.0000	.9375	.8782	.7083	.4917	.3038

TABLE 3.6: $V_0^h(\alpha^{(\ell,m)};\bar{v}^h)$ for $\sigma_1^2 = \sigma_2^2 = 10^{-5}$, $\gamma_1^2 = \gamma_2^2 = 10^{-3}$, q=0.5,$v_0$=0.2

	ℓ = 0	ℓ = 1	ℓ = 2	ℓ = 3	ℓ = 4	ℓ = 5
m = 4	.4176	.1619	.1569	.1295	.0929	.0605
m = 2	.8050	.3765	.3430	.2524	.1569	.0881
m = 0	1.0000	.8782	.8050	.6243	.4176	.2556

TABLE 3.7: $V_1^h(\alpha^{(\ell,m)};\bar{v}^h)$ for $\sigma_1^2 = \sigma_2^2 = 10^{-5}$, $\gamma_1^2 = \gamma_2^2 = 10^{-3}$, q=0.5, v_o=0.2

	ℓ = 0	ℓ = 1	ℓ = 2	ℓ = 3	ℓ = 4	ℓ = 5
m = 4	.4392	.1656	.1610	.1331	.0952	.0617
m = 2	.8560	.3947	.3594	.2626	.1610	.0886
m = 0	1.0000	.9317	.8560	.6620	.4392	.2656

Tables 3.8 and 3.9 show the values of $V_i^h(\alpha^{(\ell,o)};\bar{v}^h)$, i=0,1, respectively as functions of $v_{o,1} = v_{o,2} = v_o$.

TABLE 3.8: $V_o^h(\alpha^{(\ell,o)};\bar{v}^h)$ for $\sigma_1^2 = \sigma_2^2 = 10^{-6}$, $\gamma_1^2 = \gamma_2^2 = 10^{-4}$, q=1

	ℓ = 0	ℓ = 1	ℓ = 2	ℓ = 3	ℓ = 4	ℓ = 5
v_o = 0.1	1.0000	.9583	.7666	.4765	.2426	.1087
v_o = 0.2	1.0000	.9845	.9180	.7461	.5234	.3274
v_o = 0.3	1.0000	.9881	.9597	.8583	.6875	.4974

TABLE 3.9: $V_1^h(\alpha^{(\ell,o)};\bar{v}^h)$ for $\sigma_1^2 = \sigma_2^2 = 10^{-6}$, $\gamma_1^2 = \gamma_2^2 = 10^{-4}$, q=1

	ℓ = 0	ℓ = 1	ℓ = 2	ℓ = 3	ℓ = 4	ℓ = 5
v_o = 0.1	1.0000	.9606	.7685	.4776	.2431	.1089
v_o = 0.2	1.0000	.9873	.9208	.7484	.5249	.3283
v_o = 0.3	1.0000	.9913	.9630	.8612	.6898	.4991

Define the following functionals:

$$P_i^h(\bar{v}^h) \triangleq \int_{D_o} V_i^h(\alpha;\bar{v}^h)d\alpha / \int_{D_o} d\alpha \quad , \; i=0,1 \quad ; \qquad (3.178)$$

$$P^h_{xi}(\bar{v}^h) \triangleq \int_{D_{ox}} V^h_i(x_1,x_2,0,0;\bar{v}^h)dx/\int_{D_{ox}} dx, \; i=0,1. \tag{3.179}$$

Table 3.10 shows the values of $P^h_i(\bar{v}^h)$ and $P^h_{xi}(\bar{v}^h)$, i=0,1, as functions of q.

TABLE 3.10: $P^h_i(\bar{v}^h)$ and $P^h_{xi}(\bar{v}^h)$, i=0,1, for $\sigma_1^2 = \sigma_2^2 = 10^{-5}$, $\gamma_1^2 = \gamma_2^2 = 10^{-3}$, v_o=0.2

q	$P^h_0(\bar{v}^h)$	$P^h_1(\bar{v}^h)$	$P^h_{xo}(\bar{v}^h)$	$P^h_{x1}(\bar{v}^h)$
0.5	.0672	.0640	.2081	.2170
2.0	.0773	.0754	.2619	.2640
4.0	.0804	.0793	.2760	.2771
6.0	.0813	.0806	.2808	.2815
8.0	.0814	.0809	.2823	.2819
10.0	.0816	.0811	.2834	.2839

In order to assess the accuracy of the numerical method, the values of $V^h_i(\alpha;\bar{v}^h)$, i=0,1, were computed for h = 1/5, 1/6, 1/7. The results are given in Tables 3.11-3.12. Note that N(h), the number of points on the grid on $\bar{D}_o$ is : N(1/5) = 11^4 = 14641, N(1/6) = 13^4 = 28561 and N(1/7) = 15^4 = 50625.

TABLE 3.11: $P^h_i(\bar{v}^h)$ and $P^h_{xi}(\bar{v}^h)$, i=0,1, for $\sigma_1^2 = \sigma_2^2 = 10^{-5}$, $\gamma_1^2 = \gamma_2^2 = 10^{-3}$,q=1

	h	$P^h_0(\bar{v}^h)$	$P^h_1(\bar{v}^h)$	$P^h_{xo}(\bar{v}^h)$	$P^h_{x1}(\bar{v}^h)$
v_o = 0.1	1/5	.0624	.0584	.1684	.1699
	1/6	.0585	.0548	.1683	.1705
	1/7	.0554	.0520	.1668	.1693
v_o = 0.2	1/5	.0775	.0745	.2421	.2453
	1/6	.0733	.0705	.2414	.2456
	1/7	.0692	.0668	.2375	.2427

TABLE 3.12: $P_i^h(\bar{v}^h)$ and $P_{xi}^h(\bar{v}^h)$, i=0,1, for $\sigma_1^2 = \sigma_2^2 = 10^{-5}$, $\gamma_1^2 = \gamma_2^2 = 10^{-3}$, q=0.5 and v_o=0.15

h	$P_o^h(\bar{v}^h)$	$P_1^h(\bar{v}^h)$	$P_{xo}^h(\bar{v}^h)$	$P_{x1}^h(\bar{v}^h)$
1/5	.0695	.0638	.1927	.1977
1/6	.0649	.0598	.1902	.1969
1/7	.0607	.0561	.1854	.1935

3.8.3 Conclusions

The algorithm for computing weak suboptimal strategies, converged to the strategy $\bar{v}^h$ (eq (3.175)) in all the cases where there was convergence. In order to assess this result, the following problem is regarded.

Consider the case where the point M observes the state x(t) for all t ≥ 0. Then the motion of M is given by

$$dx_i = v_i(x)dt + \sigma_i dW_i \quad , \quad t > 0 \ , \quad i=1,2. \tag{3.180}$$

Denote

$$\Delta_T \triangleq \{x : x_1^2 + x_2^2 \le \rho^2\} \quad , \ \Delta \triangleq D_{ox} - \Delta_T \tag{3.181}$$

and let $\hat{\mathcal{D}}$ be the class of all functions V = V(x) such that V is continuous on $\bar{D}_{ox}$, twice continuously differentiable on Δ; and such that $\partial V/\partial x_i$ and $\partial^2 V/\partial x_i^2$, i=1,2, are in $L_2(D_{ox})$. Also, let $\tilde{U}_p$ be the class of all functions v = v(x), $v : \mathbb{R}^2 \to \mathbb{R}^2$ that satisfy: v_i is measurable and $|v_i(x)| \le v_o$ for all $x \in \mathbb{R}^2$, i=1,2. Given $v \in \tilde{U}_p$ and $x \in \mathbb{R}^2$ we denote by $\chi_x^v = \{\chi_x^v(t) = (\chi_{x,1}^v(t), \chi_{x,2}^v(t)),\ t \ge 0\}$ the weak solution (see Section 2.5) to eq (3.180). Define the following class of strategies

$$U_p \triangleq \{v \in \tilde{U}_p : \sup_{x \in \Delta} E_x^v \tau(x;v) < \infty\} \tag{3.182}$$

where $\tau(x;v)$ is the first exit time of $\chi_x^v(t)$ from Δ. Then, by using the methods of Yavin and Reuter [91], the following lemma can be established.

Lemma 3.3

Suppose $V_o \in \hat{\mathcal{D}}$ and $v^* \in U_p$ satisfy

$$\sum_{i=1}^{2} v_i^*(x)\partial V_o(x)/\partial x_i + (\tfrac{1}{2})\sum_{i=1}^{2} \sigma_i^2\, \partial^2 V_o(x)/\partial x_i^2 = 0 \ , \ x \in \Delta \tag{3.183}$$

$$V_o(x) = 1, \ x \in \Delta_T \quad ; \quad V_o(x) = 0 \ , \ x \notin D_{ox} \tag{3.184}$$

$$v_i^*(x) = v_o \ \mathrm{sign}(\partial V_o(x)/\partial x_i) \ , \ i=1,2, \ x \in \Delta \tag{3.185}$$

Then

$$P_x^{v^*}(\{\chi_x^{v^*}(\tau(x;v^*)) \in \Delta_T\}) \geq P_x^v(\{\chi_x^v(\tau(x;v)) \in \Delta_T\}) \tag{3.186}$$

for any $v \in U_p$ and all $x \in D_{ox}$.

Eqns (3.183)-(3.185) have been solved numerically, by using the finite-differences method described in Section 2.8, for several values of $\sigma^2 = \sigma_1^2 = \sigma_2^2$ and v_o. In all the cases considered, the optimal stra= tegy v^*, that maximizes the probability that M hits the set Δ_T before leaving D_{ox} (i.e. $P_x^v(\{\chi_x^v(\tau(x;v)) \in \Delta_T\})$), turned out to be

$$v_i^*(x) = -v_o \ \mathrm{sign}(x_i) \quad , \quad i=1,2 \quad , \quad x \in \Delta \tag{3.187}$$

where sign(o) = 0.

This result suggests that in the problem of steering M so as to hit Δ_T, by using the interrupted noisy observations of x(t), as given by (3.163), the weak suboptimal strategy $\bar{v}^h$, satisfies a 'separation principle'.

Here the phrase 'separation principle' indicates that first the optimization problem is solved by using a complete observation of x(t) and deducing an optimal strategy $v_t^* = v^*(x(t))$, and that then, for the case of interrupted noisy observation, the weak suboptimal strategy is taken to be

$$\bar{v}_i^h(y) = v_i^*(y) \quad , \ i=1,2 \quad , \ y \in D_{oy} \cap \mathbb{R}_h^2 \qquad (3.188)$$

as in our case. Here this result turned out to be valid for $q \geq 0.5$. For $q < 0.5$, the algorithm for computing weak suboptimal strategies did not converge.

Denote by T_i the (random) amount of time that Θ stays in the state i before jumping to the state j; $j \neq i$, $i,j=0,1$. It can be shown (see, for example Prohorov and Rozanov [97]) that

$$P(T_i \leq t) = 1 - e^{-qt} \quad , \ i=0,1, \ t \geq 0, \qquad (3.189)$$

and consequently

$$ET_i = \int_0^\infty t \, d_t P(T_i \leq t) = 1/q \quad , \ i=0,1 \qquad (3.190)$$

from which it follows that $q = 1/(ET_i)$, $i=0,1$.

The probabilistic meaning of q as given by (3.190) suggests, that for q small enough (which implies that ET_i will be large enough, i=0,1) a feedback strategy is no longer adequate.

3.9 EXAMPLE 3 : BANG-BANG STRATEGIES FOR HITTING A MOVING TARGET

The following example is taken from Yavin [40].

3.9.1 Statement of the Problem

Consider a random motion of two points M_p and M_e in the x_1x_2-plane. Suppose the velocity $v = (v_1,v_2)$ of M_p is perturbed by an $\mathbb{R}^2$-valued Gaussian white noise. Assume that the point M_e is moving along a line in the plane, and that its velocity $f(x_3)$, where x_3 denotes the coordinate along the line, is also perturbed by an $\mathbb{R}$-valued Gaussian white noise. Without loss of generality, the line $x_2 = 0$ may be taken as the line of motion of M_e. The observations available to the point M_p are given by

$$y = \begin{cases} (x_1,x_2,x_3) & \text{if } \theta = 1 \\ (x_1,x_2) & \text{if } \theta = 0 \end{cases} \tag{3.191}$$

where $\Theta = \{\theta(t),\ t \geq 0\}$ is a homogeneous jump Markov process with state space $S = \{0,1\}$, as described in Section 3.1.

Let D_{ox} denote the open unit square in the plane ($D_{ox} = \{x : |x_i| < 1,\ i=1,2\}$), then the problem dealt with here is to find a velocity law (strategy) $v^* = v^*(y)$ such that the probability that the point M_p hits the point M_e in D_{ox}, before either point has left D_{ox}, will be maximized on a given class of admissible strategies. (M_p hits or intercepts, M_e if M_p is in an ε-neighbourhood of M_e for a given ε).

More specifically, let (Ω,F,P) be a probability space. The motion of M_p is described by

$$dx_i = v_i(y)dt + \sigma_i dW_i \quad , \quad t > 0 \quad , \quad i=1,2 \tag{3.192}$$

and the motion of M_e given by

$$dx_3 = f(x_3)dt + \sigma_3 dW_3 \quad , \quad t > 0 \tag{3.193}$$

where $W = \{W(t) = (W_1(t),W_2(t),W_3(t)),\ t \geq 0\}$ is an $\mathbb{R}^3$-valued standard Wiener process on (Ω,F,P); σ_i, $i=1,2,3$ are given positive numbers; and f is a given continuously differentiable function.

It is assumed that the processes W and Θ are mutually independent. Owing to the nature of the observations we take the strategy v to be of the form

$$v_i(y) = \theta u_i(x) + (1-\theta)\check{v}_i(\check{x}) \quad , \quad t \geq 0 \quad , \quad i=1,2 \qquad (3.194)$$

where $x = (x_1,x_2,x_3)$ and $\check{x} = (x_1,x_2)$. It is tacitly assumed here that the process θ is observable, i.e., the point M_p knows at any time whether x or $\check{x}$ is observed.

Let $\mathcal{U}_0$, $\mathcal{U}_1$, $\tilde{U}$, u_a, $\check{v}_a$, $v_a = (u_a,\check{v}_a)$ be defined in the same manner as in Section 3.3. Also, as in Section 3.3, the following equation is considered

$$\begin{aligned} dx_i &= (\theta u_{a,i}(x) + (1-\theta)\check{v}_{a,i}(\check{x}))dt + \sigma_i dW_i, \; t > 0, \; i=1,2 \\ dx_3 &= f(x_3)dt + \sigma_3 dW_3 \; . \end{aligned} \qquad (3.195)$$

Given $v = (u,\check{v}) \in \tilde{U}$ and $x \in \mathbb{R}^3$ we denote the solution to (3.195) by $\zeta_x^v = \{\zeta_x^v(t) = (\zeta_{x,1}^v(t),\zeta_{x,2}^v(t),\zeta_{x,3}^v(t)), \; t \geq 0\}$.

Given $\varepsilon > 0$, if a certain $t \geq 0$ is the first time that $(\zeta_{x,1}^v(t) - \zeta_{x,3}^v(t))^2 + (\zeta_{x,2}^v(t))^2 \leq \varepsilon^2$, it is considered that the point M_e has been intercepted by the point M_p.

Denote by D_0, K and D the following sets in $\mathbb{R}^3$:

$$D_0 \triangleq \{x : |x_i| < 1 \quad , \quad i=1,2,3\} \qquad (3.196)$$

$$K \triangleq \{x : (x_1-x_3)^2 + x_2^2 \leq \varepsilon^2 \; , \; |x_i| \leq 1 - \delta, \; i=1,2,3\} \qquad (3.197)$$

$$D \triangleq D_0 - K \qquad (3.198)$$

where $0 < \delta << 1$ is given. If for some $t \geq 0$, $\zeta_x^v(t) \notin D$, the subsequent motion is disregarded.

Let $\tau_i(x;v)$, $i=0,1$, U and U_0 be defined by (3.12),(3.17) and (3.104) respectively and let $\mathcal{D}$ be defined as in Section 3.3. Denote

$$V_i(x;v) = P_{x,i}(\{\zeta_x^v(\tau_i(x;v)) \in K\}),\ i=0,1,\ v \in U \tag{3.15}$$

$$\ell(v) = \sum_{i=0}^{1} \int_{D_0} (1 - V_i(x;v))^2 dx\ ,\ v \in U_0. \tag{3.105}$$

In this example the following problem is treated: Find a strategy $v^o = (u^o, \check{v}^o) \in U_0$ such that $\ell(v^o) \le \ell(v)$ for any $v = (u, \check{v}) \in U_0$.

A strategy $v^o \in U_0$, whenever it exists, is supposed, roughly speaking, to steer the state $\zeta_x^{v^o}(t)$ in such a manner as to maximize, in the $L_2(D_0)$ sense, the probability that M_p intercepts M_e before either of them leaves D_0. The present example is to some extent a continuation of Yavin and Jordaan [102]. There the problem of computing optimal stra= tegies of the form $v = v(x)$ (i.e. ζ_x^v is completely observable to M_p), is considered.

The algorithm for computing weak suboptimal strategies here takes on the following form:

1. Given $v^{(n)} = (u^{(n)}, \check{v}^{(n)}) \in U_0$
2. Calculate $v_a^{(n)} = (u_a^{(n)}, \check{v}_a^{(n)})$ by using (3.47)-(3.48) (here m=3 and p=2).
3. Compute $(V_0(\cdot;v^{(n)}),\ V_1(\cdot;v^{(n)}))$ by solving numerically the follow= ing problem:

$$\sum_{i=1}^{2} \check{v}_{a,i}^{(n)}(\check{x})\partial V_0(x)/\partial x_i + f(x_3)\partial V_0(x)/\partial x_3 + (\tfrac{1}{2})\sum_{i=1}^{3} \sigma_i^2\ \partial^2 V_0(x)/\partial x_i^2 \tag{3.199}$$

$$-qV_0(x) + qV_1(x) = 0\ \ ,\ x \in D$$

$$\sum_{i=1}^{2} u_{a,i}^{(n)}(x)\partial V_1(x)/\partial x_i + f(x_3)\partial V_1(x)/\partial x_3 + (\tfrac{1}{2})\sum_{i=1}^{3} \sigma_i^2\ \partial^2 V_1(x)/\partial x_i^2 \tag{3.200}$$

$$-qV_1(x) + qV_0(x) = 0\ \ ,\ \ x \in D$$

$$V_0(x) = V_1(x) = 1\ \ ,\ x \in K;\quad V_0(x) = V_1(x) = 0\ \ ,\ \ x \notin D_0 \tag{3.201}$$

4. Calculate $\ell(v^{(n)})$

5. Compute $(Q_o(\cdot;v^{(n)}), Q_1(\cdot;v^{(n)}))$ by solving numerically the follow= ing problem:

$$-\partial[f(x_3)Q_o(x)]/\partial x_3 + (\tfrac{1}{2})\sum_{i=1}^{3}\sigma_i^2\,\partial^2 Q_o(x)/\partial x_i^2 - qQ_o(x) + qQ_1(x)$$

$$= 1 - V_o(x;v^{(n)}) \quad , \quad x \in D_o \tag{3.202}$$

$$-\partial[f(x_3)Q_1(x)]/\partial x_3 + (\tfrac{1}{2})\sum_{i=1}^{3}\sigma_i^2\,\partial^2 Q_1(x)/\partial x_i^2 - qQ_1(x) + qQ_o(x)$$

$$= 1 - V_1(x;v^{(n)}) \quad , \quad x \in D_o \tag{3.203}$$

$$Q_o(x) = Q_1(x) = 0, \quad x \notin D_o . \tag{3.104}$$

6. $v^{(n+1)}$ is determined by

$$\check{v}_i^{(n+1)}(\check{x}) = -\check{v}_{o,i}\,\mathrm{sign}\Big(\int_{D_{o\check{x}}}\theta_a^{(2)}(\check{x}-\check{x}')\int_{-1}^{1}Q_o(\check{x}',x_3;v^{(n)})(\partial V_o(\check{x}',x_3;v^{(n)})/\partial x_i)dx_3 d\check{x}\Big) \tag{3.205}$$

$i=1,2 \quad , \quad \check{x} \in D_{o\check{x}},$

$$u_i^{(n+1)}(x) = -u_{o,i}\,\mathrm{sign}\Big(\int_{D_o}\theta_a^{(3)}(x-x')Q_1(x';v^{(n)})(\partial V_1(x';v^{(n)})/\partial x_i)dx'\Big) \tag{3.206}$$

$i=1,2 \quad , \quad x \in D_o.$

7. If $v^{(n+1)} \neq v^{(n)}$; then $n + 1 \to n$, and go to 2. Otherwise: stop.

The computations are continued until for some $n \geq 0$ either $v^{(n+1)} = v^{(n)}$ or $\ell(v^{(n+1)}) = \ell(v^{(n)})$.

3.9.2 Results

The algorithm for computing weak suboptimal strategies has been applied for the following set of parameters: $\sigma_1^2 = \sigma_2^2 = 10^{-3}$, $\sigma_3^2 = 5{\cdot}10^{-3}$; $q=1$; $\varepsilon = 0.1$; $u_{o,1} = u_{o,2} = u_o = 0.05, 0.1$; $\check{v}_{o,1} = \check{v}_{o,2} = v_o = 0.025, 0.05,$

0.1; $h = 1/12$ (h is the mesh size along all axes in the finite-difference grid D_{oh} on $\mathbb{R}^3$), and $a << h$. The function f was taken as $f(x_3) = 0.01x_3, 0.05x_3, 0.1x_3$; or $f(x_3) = 0.05, 0.1$.

Denote

$$\ell^h(v^{(n)}) \triangleq h^3 \sum_{s=o}^{1} \sum_{\substack{i,j,k \\ (ih,jh,kh)\in\bar{D}_{oh}}} (1-V_s^h(ih,jh,kh;v^{(n)}))^2, n=0,1\ldots \tag{3.207}$$

In order to assess the accuracy of the numerical method the values of $\ell^h(\bar{v}^h)$ were computed for the following cases:

Case I: $u_o = 0.05$, $v_o = 0.05$, $f(x_3) = 0.05$, $q=1$;

Case II: $u_o = 0.1$, $v_o = 0.05$, $f(x_3) = 0.1$, $q=1$;

Case III: $u_o = 0.1$, $v_o = 0.025$, $f(x_3) = 0.05$, $q=2$;

Case IV: $u_o = 0.1$, $v_o = 0.05$, $f(x_3) = 0.1x_3$, $q=1$;

using $h = 1/10, 1/11, 1/12$. The results are given in Table 3.13.

TABLE 3.13: The values of $\ell^h(\bar{v}^h)$ for Case i,i=I,...,IV; for h=1/10,1/11, 1/12.

h	$\ell^h(\bar{v}^h)$:Case I	$\ell^h(\bar{v}^h)$:Case II	$\ell^h(\bar{v}^h)$:Case III	$\ell^h(\bar{v}^h)$:Case IV
1/10	9.1885	10.2878	8.4178	11.8215
1/11	8.9291	10.0810	8.1297	11.6508
1/12	8.7184	9.9182	7.8942	11.5132

In all cases solved the algorithm led either to fast convergence of $\{v^{(n)}\}$ (always for $n \le 21$), or to an oscillatory sequence $\{v^{(n)}\}$. In the second case $\bar{v}^h$ was taken as $\bar{v}^h = \arg\min_{0 \le n \le 24} \ell^h(v^{(n)})$.

The dependence of $\ell^h(\bar{v}^h)$ on q is demonstrated in Table 3.14.

TABLE 3.14: $\ell^h(\bar{v}^h)$ as function of q for $u_o = 0.1$, $v_o = 0.025$, $f(x_3) = 0.05$ and $h = 1/12$

q	0.25	0.5	1	2	4	8
$\ell^h(\bar{v}^h)$	7.9086	7.8878	7.8886	7.8942	7.8992	7.9028

For all the cases solved here the strategy $v^{(o)} = (u^{(o)},\check{v}^{(o)})$ was taken as

$$\begin{aligned} u_1^{(o)}(x) &= -u_o \operatorname{sign}(x_1-x_3) \\ u_2^{(o)}(x) &= -u_o \operatorname{sign}(x_2) \\ \check{v}_1^{(o)}(\check{x}) &= -v_o \operatorname{sign}(x_1-0.8) \\ \check{v}_2^{(o)}(\check{x}) &= -v_o \operatorname{sign}(x_2). \end{aligned} \tag{3.208}$$

The numerical computations, for all the cases solved here, led to

$$\bar{u}_2^h(x) = -\, u_o \operatorname{sign}(x_2) \quad , \tag{3.209}$$

$$\bar{\check{v}}_2^h(\check{x}) = -v_o \operatorname{sign}(x_2) \quad , \tag{3.210}$$

while $\bar{u}_1^h$ and $\bar{\check{v}}_1^h$ also followed a fixed pattern. Table 3.15 shows the values of $\bar{\check{v}}_1^h$, and Tables 3.16-3.27 show the values of $\bar{u}_1^h(x_1,x_2,k/12)$, for k = -11, -9, -7, -5, -3, -1, 1, 3, 5, 7, 9, 11 respectively for the case: $u_o = 0.1$, $v_o = 0.05$, $f(x_3) = 0.1$, $q = 1$, $h = 1/12$.

The values of $\bar{u}_1^h$ and $\bar{\check{v}}_1^h$ for other cases follow the same pattern, as shown in Tables 3.16-3.27, respectively, and thus the corresponding tables showing their values were omitted.

TABLE 3.15: The values of $\check{\bar{v}}_1^h$

.05	.05	.05	.05	.05	.05	.05	.05	.05	.05	.05	.05	.05	.05	.05	.05	.05	.05	.05	.05	-.05	-.05	-.05
.05	.05	.05	.05	.05	.05	.05	.05	.05	.05	.05	.05	.05	.05	.05	.05	.05	.05	.05	.05	-.05	-.05	-.05
.05	.05	.05	.05	.05	.05	.05	.05	.05	.05	.05	.05	.05	.05	.05	.05	.05	.05	.05	.05	-.05	-.05	-.05
.05	.05	.05	.05	.05	.05	.05	.05	.05	.05	.05	.05	.05	.05	.05	.05	.05	.05	.05	.05	-.05	-.05	-.05
.05	.05	.05	.05	.05	.05	.05	.05	.05	.05	.05	.05	.05	.05	.05	.05	.05	.05	.05	.05	.05	-.05	-.05
.05	.05	.05	.05	.05	.05	.05	.05	.05	.05	.05	.05	.05	.05	.05	.05	.05	.05	.05	.05	.05	-.05	-.05
.05	.05	.05	.05	.05	.05	.05	.05	.05	.05	.05	.05	.05	.05	.05	.05	.05	.05	.05	.05	.05	-.05	-.05
.05	.05	.05	.05	.05	.05	.05	.05	.05	.05	.05	.05	.05	.05	.05	.05	.05	.05	.05	.05	.05	-.05	-.05
.05	.05	.05	.05	.05	.05	.05	.05	.05	.05	.05	.05	.05	.05	.05	.05	.05	.05	.05	.05	.05	-.05	-.05
.05	.05	.05	.05	.05	.05	.05	.05	.05	.05	.05	.05	.05	.05	.05	.05	.05	.05	.05	.05	.05	-.05	-.05
.05	.05	.05	.05	.05	.05	.05	.05	.05	.05	.05	.05	.05	.05	.05	.05	.05	.05	.05	.05	.05	.05	-.05
.05	.05	.05	.05	.05	.05	.05	.05	.05	.05	.05	.05	.05	.05	.05	.05	.05	.05	.05	.05	.05	-.05	-.05
.05	.05	.05	.05	.05	.05	.05	.05	.05	.05	.05	.05	.05	.05	.05	.05	.05	.05	.05	.05	.05	.05	-.05
.05	.05	.05	.05	.05	.05	.05	.05	.05	.05	.05	.05	.05	.05	.05	.05	.05	.05	.05	.05	.05	-.05	-.05
.05	.05	.05	.05	.05	.05	.05	.05	.05	.05	.05	.05	.05	.05	.05	.05	.05	.05	.05	.05	.05	-.05	-.05
.05	.05	.05	.05	.05	.05	.05	.05	.05	.05	.05	.05	.05	.05	.05	.05	.05	.05	.05	.05	.05	-.05	-.05
.05	.05	.05	.05	.05	.05	.05	.05	.05	.05	.05	.05	.05	.05	.05	.05	.05	.05	.05	.05	.05	-.05	-.05
.05	.05	.05	.05	.05	.05	.05	.05	.05	.05	.05	.05	.05	.05	.05	.05	.05	.05	.05	.05	.05	-.05	-.05
.05	.05	.05	.05	.05	.05	.05	.05	.05	.05	.05	.05	.05	.05	.05	.05	.05	.05	.05	.05	.05	-.05	-.05
.05	.05	.05	.05	.05	.05	.05	.05	.05	.05	.05	.05	.05	.05	.05	.05	.05	.05	.05	.05	-.05	-.05	-.05
.05	.05	.05	.05	.05	.05	.05	.05	.05	.05	.05	.05	.05	.05	.05	.05	.05	.05	.05	.05	-.05	-.05	-.05
.05	.05	.05	.05	.05	.05	.05	.05	.05	.05	.05	.05	.05	.05	.05	.05	.05	.05	.05	.05	-.05	-.05	-.05
.05	.05	.05	.05	.05	.05	.05	.05	.05	.05	.05	.05	.05	.05	.05	.05	.05	.05	.05	.05	-.05	-.05	-.05

.1	.1	.1	.1	.1	.1	.1	.1	.1	.1	.1	.1	.1	.1	.1	.1	.1	-.1	-.1	-.1	-.1	-.1	-.1
.1	.1	.1	.1	.1	.1	.1	.1	.1	.1	.1	.1	.1	.1	.1	.1	.1	-.1	-.1	-.1	-.1	-.1	-.1
.1	.1	.1	.1	.1	.1	.1	.1	.1	.1	.1	.1	.1	.1	.1	.1	-.1	-.1	-.1	-.1	-.1	-.1	-.1
.1	.1	.1	.1	.1	.1	.1	.1	.1	.1	.1	.1	.1	.1	.1	.1	-.1	-.1	-.1	-.1	-.1	-.1	-.1
.1	.1	.1	.1	.1	.1	.1	.1	.1	.1	.1	.1	.1	.1	.1	.1	-.1	-.1	-.1	-.1	-.1	-.1	-.1
.1	.1	.1	.1	.1	.1	.1	.1	.1	.1	.1	.1	.1	.1	.1	-.1	-.1	-.1	-.1	-.1	-.1	-.1	-.1
.1	.1	.1	.1	.1	.1	.1	.1	.1	.1	.1	.1	.1	.1	.1	-.1	-.1	-.1	-.1	-.1	-.1	-.1	-.1
.1	.1	.1	.1	.1	.1	.1	.1	.1	.1	.1	.1	.1	.1	-.1	-.1	-.1	-.1	-.1	-.1	-.1	-.1	-.1
.1	.1	.1	.1	.1	.1	.1	.1	.1	.1	.1	.1	.1	-.1	-.1	-.1	-.1	-.1	-.1	-.1	-.1	-.1	-.1
.1	.1	-.1	-.1	-.1	-.1	-.1	-.1	.1	.1	.1	-.1	-.1	-.1	-.1	-.1	-.1	-.1	-.1	-.1	-.1	-.1	-.1
0.0	-.1	-.1	-.1	-.1	-.1	-.1	-.1	-.1	-.1	-.1	-.1	-.1	-.1	-.1	-.1	-.1	-.1	-.1	-.1	-.1	-.1	-.1
0.0	-.1	-.1	-.1	-.1	-.1	-.1	-.1	-.1	-.1	-.1	-.1	-.1	-.1	-.1	-.1	-.1	-.1	-.1	-.1	-.1	-.1	-.1
0.0	-.1	-.1	-.1	-.1	-.1	-.1	-.1	-.1	-.1	-.1	-.1	-.1	-.1	-.1	-.1	-.1	-.1	-.1	-.1	-.1	-.1	-.1
.1	.1	-.1	-.1	-.1	-.1	-.1	-.1	.1	.1	.1	-.1	-.1	-.1	-.1	-.1	-.1	-.1	-.1	-.1	-.1	-.1	-.1
.1	.1	.1	.1	.1	.1	.1	.1	.1	.1	.1	.1	.1	-.1	-.1	-.1	-.1	-.1	-.1	-.1	-.1	-.1	-.1
.1	.1	.1	.1	.1	.1	.1	.1	.1	.1	.1	.1	.1	.1	-.1	-.1	-.1	-.1	-.1	-.1	-.1	-.1	-.1
.1	.1	.1	.1	.1	.1	.1	.1	.1	.1	.1	.1	.1	.1	.1	-.1	-.1	-.1	-.1	-.1	-.1	-.1	-.1
.1	.1	.1	.1	.1	.1	.1	.1	.1	.1	.1	.1	.1	.1	.1	-.1	-.1	-.1	-.1	-.1	-.1	-.1	-.1
.1	.1	.1	.1	.1	.1	.1	.1	.1	.1	.1	.1	.1	.1	.1	.1	-.1	-.1	-.1	-.1	-.1	-.1	-.1
.1	.1	.1	.1	.1	.1	.1	.1	.1	.1	.1	.1	.1	.1	.1	.1	-.1	-.1	-.1	-.1	-.1	-.1	-.1
.1	.1	.1	.1	.1	.1	.1	.1	.1	.1	.1	.1	.1	.1	.1	.1	-.1	-.1	-.1	-.1	-.1	-.1	-.1
.1	.1	.1	.1	.1	.1	.1	.1	.1	.1	.1	.1	.1	.1	.1	.1	.1	-.1	-.1	-.1	-.1	-.1	-.1
.1	.1	.1	.1	.1	.1	.1	.1	.1	.1	.1	.1	.1	.1	.1	.1	.1	-.1	-.1	-.1	-.1	-.1	-.1

TABLE 3.16: The values of $\bar{u}_1^h(x_1,x_2,-11/12)$

.1	.1	.1	.1	.1	.1	.1	.1	.1	.1	.1	.1	.1	.1	.1	.1	.1	.1	-.1	-.1	-.1	-.1	-.1
.1	.1	.1	.1	.1	.1	.1	.1	.1	.1	.1	.1	.1	.1	.1	.1	.1	-.1	-.1	-.1	-.1	-.1	-.1
.1	.1	.1	.1	.1	.1	.1	.1	.1	.1	.1	.1	.1	.1	.1	.1	.1	-.1	-.1	-.1	-.1	-.1	-.1
.1	.1	.1	.1	.1	.1	.1	.1	.1	.1	.1	.1	.1	.1	.1	.1	.1	-.1	-.1	-.1	-.1	-.1	-.1
.1	.1	.1	.1	.1	.1	.1	.1	.1	.1	.1	.1	.1	.1	.1	.1	-.1	-.1	-.1	-.1	-.1	-.1	-.1
.1	.1	.1	.1	.1	.1	.1	.1	.1	.1	.1	.1	.1	.1	.1	.1	-.1	-.1	-.1	-.1	-.1	-.1	-.1
.1	.1	.1	.1	.1	.1	.1	.1	.1	.1	.1	.1	.1	.1	.1	-.1	-.1	-.1	-.1	-.1	-.1	-.1	-.1
.1	.1	.1	.1	.1	.1	.1	.1	.1	.1	.1	.1	.1	.1	.1	-.1	-.1	-.1	-.1	-.1	-.1	-.1	-.1
.1	.1	.1	.1	.1	.1	.1	.1	.1	.1	.1	.1	.1	.1	-.1	-.1	-.1	-.1	-.1	-.1	-.1	-.1	-.1
.1	.1	.1	.1	.1	.1	.1	.1	.1	.1	.1	.1	.1	-.1	-.1	-.1	-.1	-.1	-.1	-.1	-.1	-.1	-.1
.1	.1	0.0	-.1	-.1	-.1	-.1	-.1	-.1	-.1	-.1	-.1	-.1	-.1	-.1	-.1	-.1	-.1	-.1	-.1	-.1	-.1	-.1
.1	.1	0.0	-.1	-.1	-.1	-.1	-.1	-.1	-.1	-.1	-.1	-.1	-.1	-.1	-.1	-.1	-.1	-.1	-.1	-.1	-.1	-.1
.1	.1	0.0	-.1	-.1	-.1	-.1	-.1	-.1	-.1	-.1	-.1	-.1	-.1	-.1	-.1	-.1	-.1	-.1	-.1	-.1	-.1	-.1
.1	.1	.1	.1	.1	.1	.1	.1	.1	.1	.1	.1	.1	-.1	-.1	-.1	-.1	-.1	-.1	-.1	-.1	-.1	-.1
.1	.1	.1	.1	.1	.1	.1	.1	.1	.1	.1	.1	.1	.1	-.1	-.1	-.1	-.1	-.1	-.1	-.1	-.1	-.1
.1	.1	.1	.1	.1	.1	.1	.1	.1	.1	.1	.1	.1	.1	.1	-.1	-.1	-.1	-.1	-.1	-.1	-.1	-.1
.1	.1	.1	.1	.1	.1	.1	.1	.1	.1	.1	.1	.1	.1	.1	-.1	-.1	-.1	-.1	-.1	-.1	-.1	-.1
.1	.1	.1	.1	.1	.1	.1	.1	.1	.1	.1	.1	.1	.1	.1	.1	-.1	-.1	-.1	-.1	-.1	-.1	-.1
.1	.1	.1	.1	.1	.1	.1	.1	.1	.1	.1	.1	.1	.1	.1	.1	-.1	-.1	-.1	-.1	-.1	-.1	-.1
.1	.1	.1	.1	.1	.1	.1	.1	.1	.1	.1	.1	.1	.1	.1	.1	.1	-.1	-.1	-.1	-.1	-.1	-.1
.1	.1	.1	.1	.1	.1	.1	.1	.1	.1	.1	.1	.1	.1	.1	.1	.1	-.1	-.1	-.1	-.1	-.1	-.1
.1	.1	.1	.1	.1	.1	.1	.1	.1	.1	.1	.1	.1	.1	.1	.1	.1	-.1	-.1	-.1	-.1	-.1	-.1
.1	.1	.1	.1	.1	.1	.1	.1	.1	.1	.1	.1	.1	.1	.1	.1	.1	.1	-.1	-.1	-.1	-.1	-.1

TABLE 3.17: The values of $\bar{u}_1^h(x_1,x_2,-9/12)$

.1	.1	.1	.1	.1	.1	.1	.1	.1	.1	.1	.1	.1	.1	.1	.1	.1	.1	-.1	-.1	-.1	-.1	-.1
.1	.1	.1	.1	.1	.1	.1	.1	.1	.1	.1	.1	.1	.1	.1	.1	.1	.1	-.1	-.1	-.1	-.1	-.1
.1	.1	.1	.1	.1	.1	.1	.1	.1	.1	.1	.1	.1	.1	.1	.1	.1	.1	-.1	-.1	-.1	-.1	-.1
.1	.1	.1	.1	.1	.1	.1	.1	.1	.1	.1	.1	.1	.1	.1	.1	.1	.1	-.1	-.1	-.1	-.1	-.1
.1	.1	.1	.1	.1	.1	.1	.1	.1	.1	.1	.1	.1	.1	.1	.1	.1	-.1	-.1	-.1	-.1	-.1	-.1
.1	.1	.1	.1	.1	.1	.1	.1	.1	.1	.1	.1	.1	.1	.1	.1	.1	-.1	-.1	-.1	-.1	-.1	-.1
.1	.1	.1	.1	.1	.1	.1	.1	.1	.1	.1	.1	.1	.1	.1	.1	-.1	-.1	-.1	-.1	-.1	-.1	-.1
.1	.1	.1	.1	.1	.1	.1	.1	.1	.1	.1	.1	.1	.1	.1	.1	-.1	-.1	-.1	-.1	-.1	-.1	-.1
.1	.1	.1	.1	.1	.1	.1	.1	.1	.1	.1	.1	.1	.1	.1	-.1	-.1	-.1	-.1	-.1	-.1	-.1	-.1
.1	.1	.1	.1	.1	.1	.1	.1	.1	.1	.1	.1	.1	.1	-.1	-.1	-.1	-.1	-.1	-.1	-.1	-.1	-.1
.1	.1	.1	.1	0.0	-.1	-.1	-.1	.1	.1	.1	.1	-.1	-.1	-.1	-.1	-.1	-.1	-.1	-.1	-.1	-.1	-.1
.1	.1	.1	.1	0.0	-.1	-.1	-.1	-.1	-.1	-.1	-.1	-.1	-.1	-.1	-.1	-.1	-.1	-.1	-.1	-.1	-.1	-.1
.1	.1	.1	.1	0.0	-.1	-.1	-.1	.1	.1	.1	.1	-.1	-.1	-.1	-.1	-.1	-.1	-.1	-.1	-.1	-.1	-.1
.1	.1	.1	.1	.1	.1	.1	.1	.1	.1	.1	.1	.1	.1	-.1	-.1	-.1	-.1	-.1	-.1	-.1	-.1	-.1
.1	.1	.1	.1	.1	.1	.1	.1	.1	.1	.1	.1	.1	.1	.1	-.1	-.1	-.1	-.1	-.1	-.1	-.1	-.1
.1	.1	.1	.1	.1	.1	.1	.1	.1	.1	.1	.1	.1	.1	.1	.1	-.1	-.1	-.1	-.1	-.1	-.1	-.1
.1	.1	.1	.1	.1	.1	.1	.1	.1	.1	.1	.1	.1	.1	.1	.1	-.1	-.1	-.1	-.1	-.1	-.1	-.1
.1	.1	.1	.1	.1	.1	.1	.1	.1	.1	.1	.1	.1	.1	.1	.1	.1	-.1	-.1	-.1	-.1	-.1	-.1
.1	.1	.1	.1	.1	.1	.1	.1	.1	.1	.1	.1	.1	.1	.1	.1	.1	-.1	-.1	-.1	-.1	-.1	-.1
.1	.1	.1	.1	.1	.1	.1	.1	.1	.1	.1	.1	.1	.1	.1	.1	.1	.1	-.1	-.1	-.1	-.1	-.1
.1	.1	.1	.1	.1	.1	.1	.1	.1	.1	.1	.1	.1	.1	.1	.1	.1	.1	-.1	-.1	-.1	-.1	-.1
.1	.1	.1	.1	.1	.1	.1	.1	.1	.1	.1	.1	.1	.1	.1	.1	.1	.1	-.1	-.1	-.1	-.1	-.1
.1	.1	.1	.1	.1	.1	.1	.1	.1	.1	.1	.1	.1	.1	.1	.1	.1	.1	-.1	-.1	-.1	-.1	-.1

TABLE 3.18: The values of $\bar{u}_1^h(x_1,x_2,-7/12)$

.1	.1	.1	.1	.1	.1	.1	.1	.1	.1	.1	.1	.1	.1	.1	.1	.1	.1	.1	-.1	-.1	-.1	-.1
.1	.1	.1	.1	.1	.1	.1	.1	.1	.1	.1	.1	.1	.1	.1	.1	.1	.1	.1	-.1	-.1	-.1	-.1
.1	.1	.1	.1	.1	.1	.1	.1	.1	.1	.1	.1	.1	.1	.1	.1	.1	.1	.1	-.1	-.1	-.1	-.1
.1	.1	.1	.1	.1	.1	.1	.1	.1	.1	.1	.1	.1	.1	.1	.1	.1	.1	-.1	-.1	-.1	-.1	-.1
.1	.1	.1	.1	.1	.1	.1	.1	.1	.1	.1	.1	.1	.1	.1	.1	.1	.1	-.1	-.1	-.1	-.1	-.1
.1	.1	.1	.1	.1	.1	.1	.1	.1	.1	.1	.1	.1	.1	.1	.1	.1	.1	-.1	-.1	-.1	-.1	-.1
.1	.1	.1	.1	.1	.1	.1	.1	.1	.1	.1	.1	.1	.1	.1	.1	.1	-.1	-.1	-.1	-.1	-.1	-.1
.1	.1	.1	.1	.1	.1	.1	.1	.1	.1	.1	.1	.1	.1	.1	.1	.1	-.1	-.1	-.1	-.1	-.1	-.1
.1	.1	.1	.1	.1	.1	.1	.1	.1	.1	.1	.1	.1	.1	.1	.1	-.1	-.1	-.1	-.1	-.1	-.1	-.1
.1	.1	.1	.1	.1	.1	.1	.1	.1	.1	.1	.1	.1	.1	.1	-.1	-.1	-.1	-.1	-.1	-.1	-.1	-.1
.1	.1	.1	.1	.1	.1	0.0	-.1	-.1	.1	.1	.1	.1	-.1	-.1	-.1	-.1	-.1	-.1	-.1	-.1	-.1	-.1
.1	.1	.1	.1	.1	.1	0.0	-.1	-.1	-.1	-.1	-.1	-.1	-.1	-.1	-.1	-.1	-.1	-.1	-.1	-.1	-.1	-.1
.1	.1	.1	.1	.1	.1	0.0	-.1	-.1	.1	.1	.1	.1	-.1	-.1	-.1	-.1	-.1	-.1	-.1	-.1	-.1	-.1
.1	.1	.1	.1	.1	.1	.1	.1	.1	.1	.1	.1	.1	.1	.1	-.1	-.1	-.1	-.1	-.1	-.1	-.1	-.1
.1	.1	.1	.1	.1	.1	.1	.1	.1	.1	.1	.1	.1	.1	.1	.1	-.1	-.1	-.1	-.1	-.1	-.1	-.1
.1	.1	.1	.1	.1	.1	.1	.1	.1	.1	.1	.1	.1	.1	.1	.1	.1	-.1	-.1	-.1	-.1	-.1	-.1
.1	.1	.1	.1	.1	.1	.1	.1	.1	.1	.1	.1	.1	.1	.1	.1	.1	-.1	-.1	-.1	-.1	-.1	-.1
.1	.1	.1	.1	.1	.1	.1	.1	.1	.1	.1	.1	.1	.1	.1	.1	.1	.1	-.1	-.1	-.1	-.1	-.1
.1	.1	.1	.1	.1	.1	.1	.1	.1	.1	.1	.1	.1	.1	.1	.1	.1	.1	-.1	-.1	-.1	-.1	-.1
.1	.1	.1	.1	.1	.1	.1	.1	.1	.1	.1	.1	.1	.1	.1	.1	.1	.1	-.1	-.1	-.1	-.1	-.1
.1	.1	.1	.1	.1	.1	.1	.1	.1	.1	.1	.1	.1	.1	.1	.1	.1	.1	.1	-.1	-.1	-.1	-.1
.1	.1	.1	.1	.1	.1	.1	.1	.1	.1	.1	.1	.1	.1	.1	.1	.1	.1	.1	-.1	-.1	-.1	-.1
.1	.1	.1	.1	.1	.1	.1	.1	.1	.1	.1	.1	.1	.1	.1	.1	.1	.1	.1	-.1	-.1	-.1	-.1

TABLE 3.19: The values of $\bar{u}_1^h(x_1,x_2,-5/12)$

.1	.1	.1	.1	.1	.1	.1	.1	.1	.1	.1	.1	.1	.1	.1	.1	.1	.1	.1	-.1	-.1	-.1	-.1
.1	.1	.1	.1	.1	.1	.1	.1	.1	.1	.1	.1	.1	.1	.1	.1	.1	.1	.1	-.1	-.1	-.1	-.1
.1	.1	.1	.1	.1	.1	.1	.1	.1	.1	.1	.1	.1	.1	.1	.1	.1	.1	.1	-.1	-.1	-.1	-.1
.1	.1	.1	.1	.1	.1	.1	.1	.1	.1	.1	.1	.1	.1	.1	.1	.1	.1	.1	-.1	-.1	-.1	-.1
.1	.1	.1	.1	.1	.1	.1	.1	.1	.1	.1	.1	.1	.1	.1	.1	.1	.1	.1	-.1	-.1	-.1	-.1
.1	.1	.1	.1	.1	.1	.1	.1	.1	.1	.1	.1	.1	.1	.1	.1	.1	.1	.1	-.1	-.1	-.1	-.1
.1	.1	.1	.1	.1	.1	.1	.1	.1	.1	.1	.1	.1	.1	.1	.1	.1	.1	-.1	-.1	-.1	-.1	-.1
.1	.1	.1	.1	.1	.1	.1	.1	.1	.1	.1	.1	.1	.1	.1	.1	.1	.1	-.1	-.1	-.1	-.1	-.1
.1	.1	.1	.1	.1	.1	.1	.1	.1	.1	.1	.1	.1	.1	.1	.1	.1	-.1	-.1	-.1	-.1	-.1	-.1
.1	.1	.1	.1	.1	.1	.1	.1	.1	.1	.1	.1	.1	.1	.1	.1	-.1	-.1	-.1	-.1	-.1	-.1	-.1
.1	.1	.1	.1	.1	.1	.1	.1	0.0	-.1	-.1	.1	.1	.1	-.1	-.1	-.1	-.1	-.1	-.1	-.1	-.1	-.1
.1	.1	.1	.1	.1	.1	.1	.1	0.0	-.1	-.1	-.1	-.1	-.1	-.1	-.1	-.1	-.1	-.1	-.1	-.1	-.1	-.1
.1	.1	.1	.1	.1	.1	.1	.1	0.0	-.1	-.1	.1	.1	.1	-.1	-.1	-.1	-.1	-.1	-.1	-.1	-.1	-.1
.1	.1	.1	.1	.1	.1	.1	.1	.1	.1	.1	.1	.1	.1	.1	.1	-.1	-.1	-.1	-.1	-.1	-.1	-.1
.1	.1	.1	.1	.1	.1	.1	.1	.1	.1	.1	.1	.1	.1	.1	.1	.1	-.1	-.1	-.1	-.1	-.1	-.1
.1	.1	.1	.1	.1	.1	.1	.1	.1	.1	.1	.1	.1	.1	.1	.1	.1	.1	-.1	-.1	-.1	-.1	-.1
.1	.1	.1	.1	.1	.1	.1	.1	.1	.1	.1	.1	.1	.1	.1	.1	.1	.1	-.1	-.1	-.1	-.1	-.1
.1	.1	.1	.1	.1	.1	.1	.1	.1	.1	.1	.1	.1	.1	.1	.1	.1	.1	.1	-.1	-.1	-.1	-.1
.1	.1	.1	.1	.1	.1	.1	.1	.1	.1	.1	.1	.1	.1	.1	.1	.1	.1	.1	-.1	-.1	-.1	-.1
.1	.1	.1	.1	.1	.1	.1	.1	.1	.1	.1	.1	.1	.1	.1	.1	.1	.1	.1	-.1	-.1	-.1	-.1
.1	.1	.1	.1	.1	.1	.1	.1	.1	.1	.1	.1	.1	.1	.1	.1	.1	.1	.1	-.1	-.1	-.1	-.1
.1	.1	.1	.1	.1	.1	.1	.1	.1	.1	.1	.1	.1	.1	.1	.1	.1	.1	.1	-.1	-.1	-.1	-.1
.1	.1	.1	.1	.1	.1	.1	.1	.1	.1	.1	.1	.1	.1	.1	.1	.1	.1	.1	-.1	-.1	-.1	-.1

TABLE 3.20: The values of $\bar{u}_1^h(x_1,x_2,-3/12)$

.1	.1	.1	.1	.1	.1	.1	.1	.1	.1	.1	.1	.1	.1	.1	.1	.1	.1	.1	.1	-.1	-.1	-.1
.1	.1	.1	.1	.1	.1	.1	.1	.1	.1	.1	.1	.1	.1	.1	.1	.1	.1	.1	.1	-.1	-.1	-.1
.1	.1	.1	.1	.1	.1	.1	.1	.1	.1	.1	.1	.1	.1	.1	.1	.1	.1	.1	.1	-.1	-.1	-.1
.1	.1	.1	.1	.1	.1	.1	.1	.1	.1	.1	.1	.1	.1	.1	.1	.1	.1	.1	.1	-.1	-.1	-.1
.1	.1	.1	.1	.1	.1	.1	.1	.1	.1	.1	.1	.1	.1	.1	.1	.1	.1	.1	-.1	-.1	-.1	-.1
.1	.1	.1	.1	.1	.1	.1	.1	.1	.1	.1	.1	.1	.1	.1	.1	.1	.1	.1	-.1	-.1	-.1	-.1
.1	.1	.1	.1	.1	.1	.1	.1	.1	.1	.1	.1	.1	.1	.1	.1	.1	.1	.1	-.1	-.1	-.1	-.1
.1	.1	.1	.1	.1	.1	.1	.1	.1	.1	.1	.1	.1	.1	.1	.1	.1	.1	.1	-.1	-.1	-.1	-.1
.1	.1	.1	.1	.1	.1	.1	.1	.1	.1	.1	.1	.1	.1	.1	.1	.1	.1	-.1	-.1	-.1	-.1	-.1
.1	.1	.1	.1	.1	.1	.1	.1	.1	.1	.1	.1	.1	.1	.1	.1	.1	-.1	-.1	-.1	-.1	-.1	-.1
.1	.1	.1	.1	.1	.1	.1	.1	.1	.1	0.0	-.1	-.1	-.1	-.1	-.1	-.1	-.1	-.1	-.1	-.1	-.1	-.1
.1	.1	.1	.1	.1	.1	.1	.1	.1	.1	0.0	-.1	-.1	-.1	-.1	-.1	-.1	-.1	-.1	-.1	-.1	-.1	-.1
.1	.1	.1	.1	.1	.1	.1	.1	.1	.1	0.0	-.1	-.1	-.1	-.1	-.1	-.1	-.1	-.1	-.1	-.1	-.1	-.1
.1	.1	.1	.1	.1	.1	.1	.1	.1	.1	.1	.1	.1	.1	.1	.1	.1	-.1	-.1	-.1	-.1	-.1	-.1
.1	.1	.1	.1	.1	.1	.1	.1	.1	.1	.1	.1	.1	.1	.1	.1	.1	.1	-.1	-.1	-.1	-.1	-.1
.1	.1	.1	.1	.1	.1	.1	.1	.1	.1	.1	.1	.1	.1	.1	.1	.1	.1	.1	-.1	-.1	-.1	-.1
.1	.1	.1	.1	.1	.1	.1	.1	.1	.1	.1	.1	.1	.1	.1	.1	.1	.1	.1	-.1	-.1	-.1	-.1
.1	.1	.1	.1	.1	.1	.1	.1	.1	.1	.1	.1	.1	.1	.1	.1	.1	.1	.1	-.1	-.1	-.1	-.1
.1	.1	.1	.1	.1	.1	.1	.1	.1	.1	.1	.1	.1	.1	.1	.1	.1	.1	.1	-.1	-.1	-.1	-.1
.1	.1	.1	.1	.1	.1	.1	.1	.1	.1	.1	.1	.1	.1	.1	.1	.1	.1	.1	.1	-.1	-.1	-.1
.1	.1	.1	.1	.1	.1	.1	.1	.1	.1	.1	.1	.1	.1	.1	.1	.1	.1	.1	.1	-.1	-.1	-.1
.1	.1	.1	.1	.1	.1	.1	.1	.1	.1	.1	.1	.1	.1	.1	.1	.1	.1	.1	.1	-.1	-.1	-.1
.1	.1	.1	.1	.1	.1	.1	.1	.1	.1	.1	.1	.1	.1	.1	.1	.1	.1	.1	.1	-.1	-.1	-.1

TABLE 3.21: The values of $\bar{u}_1^h(x_1,x_2,-1/12)$

.1	.1	.1	.1	.1	.1	.1	.1	.1	.1	.1	.1	.1	.1	.1	.1	.1	.1	.1	.1	-.1	-.1	-.1
.1	.1	.1	.1	.1	.1	.1	.1	.1	.1	.1	.1	.1	.1	.1	.1	.1	.1	.1	.1	-.1	-.1	-.1
.1	.1	.1	.1	.1	.1	.1	.1	.1	.1	.1	.1	.1	.1	.1	.1	.1	.1	.1	.1	-.1	-.1	-.1
.1	.1	.1	.1	.1	.1	.1	.1	.1	.1	.1	.1	.1	.1	.1	.1	.1	.1	.1	.1	-.1	-.1	-.1
.1	.1	.1	.1	.1	.1	.1	.1	.1	.1	.1	.1	.1	.1	.1	.1	.1	.1	.1	.1	-.1	-.1	-.1
.1	.1	.1	.1	.1	.1	.1	.1	.1	.1	.1	.1	.1	.1	.1	.1	.1	.1	.1	.1	-.1	-.1	-.1
.1	.1	.1	.1	.1	.1	.1	.1	.1	.1	.1	.1	.1	.1	.1	.1	.1	.1	.1	.1	-.1	-.1	-.1
.1	.1	.1	.1	.1	.1	.1	.1	.1	.1	.1	.1	.1	.1	.1	.1	.1	.1	.1	.1	-.1	-.1	-.1
.1	.1	.1	.1	.1	.1	.1	.1	.1	.1	.1	.1	.1	.1	.1	.1	.1	.1	.1	-.1	-.1	-.1	-.1
.1	.1	.1	.1	.1	.1	.1	.1	.1	.1	.1	.1	.1	.1	.1	.1	.1	.1	-.1	-.1	-.1	-.1	-.1
.1	.1	.1	.1	.1	.1	.1	.1	.1	.1	.1	.1	0.0	-.1	-.1	-.1	-.1	-.1	-.1	-.1	-.1	-.1	-.1
.1	.1	.1	.1	.1	.1	.1	.1	.1	.1	.1	.1	0.0	-.1	-.1	-.1	-.1	-.1	-.1	-.1	-.1	-.1	-.1
.1	.1	.1	.1	.1	.1	.1	.1	.1	.1	.1	.1	0.0	-.1	-.1	-.1	-.1	-.1	-.1	-.1	-.1	-.1	-.1
.1	.1	.1	.1	.1	.1	.1	.1	.1	.1	.1	.1	.1	.1	.1	.1	.1	.1	-.1	-.1	-.1	-.1	-.1
.1	.1	.1	.1	.1	.1	.1	.1	.1	.1	.1	.1	.1	.1	.1	.1	.1	.1	.1	-.1	-.1	-.1	-.1
.1	.1	.1	.1	.1	.1	.1	.1	.1	.1	.1	.1	.1	.1	.1	.1	.1	.1	.1	.1	-.1	-.1	-.1
.1	.1	.1	.1	.1	.1	.1	.1	.1	.1	.1	.1	.1	.1	.1	.1	.1	.1	.1	.1	-.1	-.1	-.1
.1	.1	.1	.1	.1	.1	.1	.1	.1	.1	.1	.1	.1	.1	.1	.1	.1	.1	.1	.1	-.1	-.1	-.1
.1	.1	.1	.1	.1	.1	.1	.1	.1	.1	.1	.1	.1	.1	.1	.1	.1	.1	.1	.1	-.1	-.1	-.1
.1	.1	.1	.1	.1	.1	.1	.1	.1	.1	.1	.1	.1	.1	.1	.1	.1	.1	.1	.1	-.1	-.1	-.1
.1	.1	.1	.1	.1	.1	.1	.1	.1	.1	.1	.1	.1	.1	.1	.1	.1	.1	.1	.1	-.1	-.1	-.1
.1	.1	.1	.1	.1	.1	.1	.1	.1	.1	.1	.1	.1	.1	.1	.1	.1	.1	.1	.1	-.1	-.1	-.1
.1	.1	.1	.1	.1	.1	.1	.1	.1	.1	.1	.1	.1	.1	.1	.1	.1	.1	.1	.1	-.1	-.1	-.1

TABLE 3.22: The values of $\bar{u}_1^h(x_1,x_2,1/12)$

.1	.1	.1	.1	.1	.1	.1	.1	.1	.1	.1	.1	.1	.1	.1	.1	.1	.1	.1	.1	-.1	-.1	-.1
.1	.1	.1	.1	.1	.1	.1	.1	.1	.1	.1	.1	.1	.1	.1	.1	.1	.1	.1	.1	-.1	-.1	-.1
.1	.1	.1	.1	.1	.1	.1	.1	.1	.1	.1	.1	.1	.1	.1	.1	.1	.1	.1	.1	-.1	-.1	-.1
.1	.1	.1	.1	.1	.1	.1	.1	.1	.1	.1	.1	.1	.1	.1	.1	.1	.1	.1	.1	-.1	-.1	-.1
.1	.1	.1	.1	.1	.1	.1	.1	.1	.1	.1	.1	.1	.1	.1	.1	.1	.1	.1	.1	-.1	-.1	-.1
.1	.1	.1	.1	.1	.1	.1	.1	.1	.1	.1	.1	.1	.1	.1	.1	.1	.1	.1	.1	-.1	-.1	-.1
.1	.1	.1	.1	.1	.1	.1	.1	.1	.1	.1	.1	.1	.1	.1	.1	.1	.1	.1	.1	-.1	-.1	-.1
.1	.1	.1	.1	.1	.1	.1	.1	.1	.1	.1	.1	.1	.1	.1	.1	.1	.1	.1	.1	-.1	-.1	-.1
.1	.1	.1	.1	.1	.1	.1	.1	.1	.1	.1	.1	.1	.1	.1	.1	.1	.1	.1	.1	-.1	-.1	-.1
.1	.1	.1	.1	.1	.1	.1	.1	.1	.1	.1	.1	.1	.1	.1	.1	.1	.1	.1	-.1	-.1	-.1	-.1
.1	.1	.1	.1	.1	.1	.1	.1	.1	.1	.1	.1	.1	.1	0.0	-.1	-.1	-.1	-.1	-.1	-.1	-.1	-.1
.1	.1	.1	.1	.1	.1	.1	.1	.1	.1	.1	.1	.1	.1	0.0	-.1	-.1	-.1	-.1	-.1	-.1	-.1	-.1
.1	.1	.1	.1	.1	.1	.1	.1	.1	.1	.1	.1	.1	.1	0.0	-.1	-.1	-.1	-.1	-.1	-.1	-.1	-.1
.1	.1	.1	.1	.1	.1	.1	.1	.1	.1	.1	.1	.1	.1	.1	.1	.1	.1	.1	-.1	-.1	-.1	-.1
.1	.1	.1	.1	.1	.1	.1	.1	.1	.1	.1	.1	.1	.1	.1	.1	.1	.1	.1	.1	-.1	-.1	-.1
.1	.1	.1	.1	.1	.1	.1	.1	.1	.1	.1	.1	.1	.1	.1	.1	.1	.1	.1	.1	-.1	-.1	-.1
.1	.1	.1	.1	.1	.1	.1	.1	.1	.1	.1	.1	.1	.1	.1	.1	.1	.1	.1	.1	-.1	-.1	-.1
.1	.1	.1	.1	.1	.1	.1	.1	.1	.1	.1	.1	.1	.1	.1	.1	.1	.1	.1	.1	-.1	-.1	-.1
.1	.1	.1	.1	.1	.1	.1	.1	.1	.1	.1	.1	.1	.1	.1	.1	.1	.1	.1	.1	-.1	-.1	-.1
.1	.1	.1	.1	.1	.1	.1	.1	.1	.1	.1	.1	.1	.1	.1	.1	.1	.1	.1	.1	-.1	-.1	-.1
.1	.1	.1	.1	.1	.1	.1	.1	.1	.1	.1	.1	.1	.1	.1	.1	.1	.1	.1	.1	-.1	-.1	-.1
.1	.1	.1	.1	.1	.1	.1	.1	.1	.1	.1	.1	.1	.1	.1	.1	.1	.1	.1	.1	-.1	-.1	-.1
.1	.1	.1	.1	.1	.1	.1	.1	.1	.1	.1	.1	.1	.1	.1	.1	.1	.1	.1	.1	-.1	-.1	-.1

TABLE 3.23: The values of $\bar{u}_1(x_1,x_2,3/12)$

.1	.1	.1	.1	.1	.1	.1	.1	.1	.1	.1	.1	.1	.1	.1	.1	.1	.1	.1	.1	-.1	-.1	-.1
.1	.1	.1	.1	.1	.1	.1	.1	.1	.1	.1	.1	.1	.1	.1	.1	.1	.1	.1	.1	-.1	-.1	-.1
.1	.1	.1	.1	.1	.1	.1	.1	.1	.1	.1	.1	.1	.1	.1	.1	.1	.1	.1	.1	.1	-.1	-.1
.1	.1	.1	.1	.1	.1	.1	.1	.1	.1	.1	.1	.1	.1	.1	.1	.1	.1	.1	.1	.1	-.1	-.1
.1	.1	.1	.1	.1	.1	.1	.1	.1	.1	.1	.1	.1	.1	.1	.1	.1	.1	.1	.1	.1	-.1	-.1
.1	.1	.1	.1	.1	.1	.1	.1	.1	.1	.1	.1	.1	.1	.1	.1	.1	.1	.1	.1	.1	-.1	-.1
.1	.1	.1	.1	.1	.1	.1	.1	.1	.1	.1	.1	.1	.1	.1	.1	.1	.1	.1	.1	.1	-.1	-.1
.1	.1	.1	.1	.1	.1	.1	.1	.1	.1	.1	.1	.1	.1	.1	.1	.1	.1	.1	.1	.1	-.1	-.1
.1	.1	.1	.1	.1	.1	.1	.1	.1	.1	.1	.1	.1	.1	.1	.1	.1	.1	.1	.1	.1	-.1	-.1
.1	.1	.1	.1	.1	.1	.1	.1	.1	.1	.1	.1	.1	.1	.1	.1	.1	.1	.1	.1	-.1	-.1	-.1
.1	.1	.1	.1	.1	.1	.1	.1	.1	.1	.1	.1	.1	.1	.1	.1	0.0	-.1	-.1	-.1	-.1	-.1	-.1
.1	.1	.1	.1	.1	.1	.1	.1	.1	.1	.1	.1	.1	.1	.1	.1	0.0	-.1	-.1	-.1	-.1	-.1	-.1
.1	.1	.1	.1	.1	.1	.1	.1	.1	.1	.1	.1	.1	.1	.1	.1	0.0	-.1	-.1	-.1	-.1	-.1	-.1
.1	.1	.1	.1	.1	.1	.1	.1	.1	.1	.1	.1	.1	.1	.1	.1	.1	.1	.1	.1	-.1	-.1	-.1
.1	.1	.1	.1	.1	.1	.1	.1	.1	.1	.1	.1	.1	.1	.1	.1	.1	.1	.1	.1	.1	-.1	-.1
.1	.1	.1	.1	.1	.1	.1	.1	.1	.1	.1	.1	.1	.1	.1	.1	.1	.1	.1	.1	.1	-.1	-.1
.1	.1	.1	.1	.1	.1	.1	.1	.1	.1	.1	.1	.1	.1	.1	.1	.1	.1	.1	.1	.1	-.1	-.1
.1	.1	.1	.1	.1	.1	.1	.1	.1	.1	.1	.1	.1	.1	.1	.1	.1	.1	.1	.1	.1	-.1	-.1
.1	.1	.1	.1	.1	.1	.1	.1	.1	.1	.1	.1	.1	.1	.1	.1	.1	.1	.1	.1	.1	-.1	-.1
.1	.1	.1	.1	.1	.1	.1	.1	.1	.1	.1	.1	.1	.1	.1	.1	.1	.1	.1	.1	.1	-.1	-.1
.1	.1	.1	.1	.1	.1	.1	.1	.1	.1	.1	.1	.1	.1	.1	.1	.1	.1	.1	.1	.1	-.1	-.1
.1	.1	.1	.1	.1	.1	.1	.1	.1	.1	.1	.1	.1	.1	.1	.1	.1	.1	.1	.1	-.1	-.1	-.1
.1	.1	.1	.1	.1	.1	.1	.1	.1	.1	.1	.1	.1	.1	.1	.1	.1	.1	.1	.1	-.1	-.1	-.1

TABLE 3.24: The values of $\bar{u}_1^h(x_1,x_2,5/12)$

.1	.1	.1	.1	.1	.1	.1	.1	.1	.1	.1	.1	.1	.1	.1	.1	.1	.1	.1	.1	.1	-.1	-.1
.1	.1	.1	.1	.1	.1	.1	.1	.1	.1	.1	.1	.1	.1	.1	.1	.1	.1	.1	.1	.1	-.1	-.1
.1	.1	.1	.1	.1	.1	.1	.1	.1	.1	.1	.1	.1	.1	.1	.1	.1	.1	.1	.1	.1	-.1	-.1
.1	.1	.1	.1	.1	.1	.1	.1	.1	.1	.1	.1	.1	.1	.1	.1	.1	.1	.1	.1	.1	-.1	-.1
.1	.1	.1	.1	.1	.1	.1	.1	.1	.1	.1	.1	.1	.1	.1	.1	.1	.1	.1	.1	.1	-.1	-.1
.1	.1	.1	.1	.1	.1	.1	.1	.1	.1	.1	.1	.1	.1	.1	.1	.1	.1	.1	.1	.1	-.1	-.1
.1	.1	.1	.1	.1	.1	.1	.1	.1	.1	.1	.1	.1	.1	.1	.1	.1	.1	.1	.1	.1	-.1	-.1
.1	.1	.1	.1	.1	.1	.1	.1	.1	.1	.1	.1	.1	.1	.1	.1	.1	.1	.1	.1	.1	-.1	-.1
.1	.1	.1	.1	.1	.1	.1	.1	.1	.1	.1	.1	.1	.1	.1	.1	.1	.1	.1	.1	.1	-.1	-.1
.1	.1	.1	.1	.1	.1	.1	.1	.1	.1	.1	.1	.1	.1	.1	.1	.1	.1	.1	.1	.1	-.1	-.1
.1	.1	.1	.1	.1	.1	.1	.1	.1	.1	.1	.1	.1	.1	.1	.1	.1	.1	0.0	-.1	-.1	-.1	-.1
.1	.1	.1	.1	.1	.1	.1	.1	.1	.1	.1	.1	.1	.1	.1	.1	.1	.1	0.0	-.1	-.1	-.1	-.1
.1	.1	.1	.1	.1	.1	.1	.1	.1	.1	.1	.1	.1	.1	.1	.1	.1	.1	0.0	-.1	-.1	-.1	-.1
.1	.1	.1	.1	.1	.1	.1	.1	.1	.1	.1	.1	.1	.1	.1	.1	.1	.1	.1	.1	.1	-.1	-.1
.1	.1	.1	.1	.1	.1	.1	.1	.1	.1	.1	.1	.1	.1	.1	.1	.1	.1	.1	.1	.1	-.1	-.1
.1	.1	.1	.1	.1	.1	.1	.1	.1	.1	.1	.1	.1	.1	.1	.1	.1	.1	.1	.1	.1	-.1	-.1
.1	.1	.1	.1	.1	.1	.1	.1	.1	.1	.1	.1	.1	.1	.1	.1	.1	.1	.1	.1	.1	-.1	-.1
.1	.1	.1	.1	.1	.1	.1	.1	.1	.1	.1	.1	.1	.1	.1	.1	.1	.1	.1	.1	.1	-.1	-.1
.1	.1	.1	.1	.1	.1	.1	.1	.1	.1	.1	.1	.1	.1	.1	.1	.1	.1	.1	.1	.1	-.1	-.1
.1	.1	.1	.1	.1	.1	.1	.1	.1	.1	.1	.1	.1	.1	.1	.1	.1	.1	.1	.1	.1	-.1	-.1
.1	.1	.1	.1	.1	.1	.1	.1	.1	.1	.1	.1	.1	.1	.1	.1	.1	.1	.1	.1	.1	-.1	-.1
.1	.1	.1	.1	.1	.1	.1	.1	.1	.1	.1	.1	.1	.1	.1	.1	.1	.1	.1	.1	.1	-.1	-.1
.1	.1	.1	.1	.1	.1	.1	.1	.1	.1	.1	.1	.1	.1	.1	.1	.1	.1	.1	.1	.1	-.1	-.1

TABLE 3.25: The values of $\bar{u}_1^h(x_1,x_2,7/12)$

.1	.1	.1	.1	.1	.1	.1	.1	.1	.1	.1	.1	.1	.1	.1	.1	.1	.1	.1	.1	.1	-.1	-.1
.1	.1	.1	.1	.1	.1	.1	.1	.1	.1	.1	.1	.1	.1	.1	.1	.1	.1	.1	.1	.1	-.1	-.1
.1	.1	.1	.1	.1	.1	.1	.1	.1	.1	.1	.1	.1	.1	.1	.1	.1	.1	.1	.1	.1	-.1	-.1
.1	.1	.1	.1	.1	.1	.1	.1	.1	.1	.1	.1	.1	.1	.1	.1	.1	.1	.1	.1	.1	-.1	-.1
.1	.1	.1	.1	.1	.1	.1	.1	.1	.1	.1	.1	.1	.1	.1	.1	.1	.1	.1	.1	.1	-.1	-.1
.1	.1	.1	.1	.1	.1	.1	.1	.1	.1	.1	.1	.1	.1	.1	.1	.1	.1	.1	.1	.1	-.1	-.1
.1	.1	.1	.1	.1	.1	.1	.1	.1	.1	.1	.1	.1	.1	.1	.1	.1	.1	.1	.1	.1	-.1	-.1
.1	.1	.1	.1	.1	.1	.1	.1	.1	.1	.1	.1	.1	.1	.1	.1	.1	.1	.1	.1	.1	.1	-.1
.1	.1	.1	.1	.1	.1	.1	.1	.1	.1	.1	.1	.1	.1	.1	.1	.1	.1	.1	.1	.1	.1	-.1
.1	.1	.1	.1	.1	.1	.1	.1	.1	.1	.1	.1	.1	.1	.1	.1	.1	.1	.1	.1	.1	.1	-.1
.1	.1	.1	.1	.1	.1	.1	.1	.1	.1	.1	.1	.1	.1	.1	.1	.1	.1	.1	.1	0.0	-.1	-.1
.1	.1	.1	.1	.1	.1	.1	.1	.1	.1	.1	.1	.1	.1	.1	.1	.1	.1	.1	.1	0.0	-.1	-.1
.1	.1	.1	.1	.1	.1	.1	.1	.1	.1	.1	.1	.1	.1	.1	.1	.1	.1	.1	.1	0.0	-.1	-.1
.1	.1	.1	.1	.1	.1	.1	.1	.1	.1	.1	.1	.1	.1	.1	.1	.1	.1	.1	.1	.1	.1	-.1
.1	.1	.1	.1	.1	.1	.1	.1	.1	.1	.1	.1	.1	.1	.1	.1	.1	.1	.1	.1	.1	.1	-.1
.1	.1	.1	.1	.1	.1	.1	.1	.1	.1	.1	.1	.1	.1	.1	.1	.1	.1	.1	.1	.1	.1	-.1
.1	.1	.1	.1	.1	.1	.1	.1	.1	.1	.1	.1	.1	.1	.1	.1	.1	.1	.1	.1	.1	-.1	-.1
.1	.1	.1	.1	.1	.1	.1	.1	.1	.1	.1	.1	.1	.1	.1	.1	.1	.1	.1	.1	.1	-.1	-.1
.1	.1	.1	.1	.1	.1	.1	.1	.1	.1	.1	.1	.1	.1	.1	.1	.1	.1	.1	.1	.1	-.1	-.1
.1	.1	.1	.1	.1	.1	.1	.1	.1	.1	.1	.1	.1	.1	.1	.1	.1	.1	.1	.1	.1	-.1	-.1
.1	.1	.1	.1	.1	.1	.1	.1	.1	.1	.1	.1	.1	.1	.1	.1	.1	.1	.1	.1	.1	-.1	-.1
.1	.1	.1	.1	.1	.1	.1	.1	.1	.1	.1	.1	.1	.1	.1	.1	.1	.1	.1	.1	.1	-.1	-.1
.1	.1	.1	.1	.1	.1	.1	.1	.1	.1	.1	.1	.1	.1	.1	.1	.1	.1	.1	.1	.1	-.1	-.1

TABLE 3.26: The values of $\bar{u}_1^h(x_1,x_2,9/12)$

.1	.1	.1	.1	.1	.1	.1	.1	.1	.1	.1	.1	.1	.1	.1	.1	.1	.1	.1	.1	.1	-.1	-.1
.1	.1	.1	.1	.1	.1	.1	.1	.1	.1	.1	.1	.1	.1	.1	.1	.1	.1	.1	.1	.1	-.1	-.1
.1	.1	.1	.1	.1	.1	.1	.1	.1	.1	.1	.1	.1	.1	.1	.1	.1	.1	.1	.1	.1	-.1	-.1
.1	.1	.1	.1	.1	.1	.1	.1	.1	.1	.1	.1	.1	.1	.1	.1	.1	.1	.1	.1	.1	-.1	-.1
.1	.1	.1	.1	.1	.1	.1	.1	.1	.1	.1	.1	.1	.1	.1	.1	.1	.1	.1	.1	.1	-.1	-.1
.1	.1	.1	.1	.1	.1	.1	.1	.1	.1	.1	.1	.1	.1	.1	.1	.1	.1	.1	.1	.1	.1	-.1
.1	.1	.1	.1	.1	.1	.1	.1	.1	.1	.1	.1	.1	.1	.1	.1	.1	.1	.1	.1	.1	.1	-.1
.1	.1	.1	.1	.1	.1	.1	.1	.1	.1	.1	.1	.1	.1	.1	.1	.1	.1	.1	.1	.1	.1	-.1
.1	.1	.1	.1	.1	.1	.1	.1	.1	.1	.1	.1	.1	.1	.1	.1	.1	.1	.1	.1	.1	.1	-.1
.1	.1	.1	.1	.1	.1	.1	.1	.1	.1	.1	.1	.1	.1	.1	.1	.1	.1	.1	.1	.1	.1	-.1
.1	.1	.1	.1	.1	.1	.1	.1	.1	.1	.1	.1	.1	.1	.1	.1	.1	.1	.1	.1	.1	.1	0.0
.1	.1	.1	.1	.1	.1	.1	.1	.1	.1	.1	.1	.1	.1	.1	.1	.1	.1	.1	.1	.1	.1	0.0
.1	.1	.1	.1	.1	.1	.1	.1	.1	.1	.1	.1	.1	.1	.1	.1	.1	.1	.1	.1	.1	.1	0.0
.1	.1	.1	.1	.1	.1	.1	.1	.1	.1	.1	.1	.1	.1	.1	.1	.1	.1	.1	.1	.1	.1	-.1
.1	.1	.1	.1	.1	.1	.1	.1	.1	.1	.1	.1	.1	.1	.1	.1	.1	.1	.1	.1	.1	.1	-.1
.1	.1	.1	.1	.1	.1	.1	.1	.1	.1	.1	.1	.1	.1	.1	.1	.1	.1	.1	.1	.1	.1	-.1
.1	.1	.1	.1	.1	.1	.1	.1	.1	.1	.1	.1	.1	.1	.1	.1	.1	.1	.1	.1	.1	.1	-.1
.1	.1	.1	.1	.1	.1	.1	.1	.1	.1	.1	.1	.1	.1	.1	.1	.1	.1	.1	.1	.1	-.1	-.1
.1	.1	.1	.1	.1	.1	.1	.1	.1	.1	.1	.1	.1	.1	.1	.1	.1	.1	.1	.1	.1	-.1	-.1
.1	.1	.1	.1	.1	.1	.1	.1	.1	.1	.1	.1	.1	.1	.1	.1	.1	.1	.1	.1	.1	-.1	-.1
.1	.1	.1	.1	.1	.1	.1	.1	.1	.1	.1	.1	.1	.1	.1	.1	.1	.1	.1	.1	.1	-.1	-.1
.1	.1	.1	.1	.1	.1	.1	.1	.1	.1	.1	.1	.1	.1	.1	.1	.1	.1	.1	.1	.1	-.1	-.1
.1	.1	.1	.1	.1	.1	.1	.1	.1	.1	.1	.1	.1	.1	.1	.1	.1	.1	.1	.1	.1	-.1	-.1

TABLE 3.27: The values of $\bar{u}_1^h(x_1,x_2,11/12)$

Denote

$$V_i^h(x_1,x_2;\bar{v}^h) \triangleq \int_{-1}^{1} V_i^h(x_1,x_2,x_3;\bar{v}^h)dx_3 / \int_{-1}^{1} dx_3$$

$$= V_i^h(\check{x};\bar{v}^h) \quad , \quad i=0,1. \tag{3.211}$$

The values of $V_i(x) = V_i^h(x;\bar{v}^h)$ and $V_i(\check{x}) = V_i(\check{x};\bar{v}^h)$, i=0,1, are shown in Figures 3.5-3.11 for two cases:

Case 1 : $u_o = 0.05$, $v_o = 0.05$, $f(x_3) = 0.05$, $q = 1$, $h = 1/12$;

Case 2 : $u_o = 0.1$, $v_o = 0.1$, $f(x_3) = 0.05$, $q = 1$, $h = 1/12$.

Figures 3.5-3.7 show the results obtained for Case 1, and Figures 3.8-3.11 show the results obtained for Case 2.

In all the cases computed it turned out as expected that

$$V_i^h(x_1,-x_2,x_3;\bar{v}^h) = V_i^h(x_1,x_2,x_3;\bar{v}^h) \quad , \quad i=0,1 \quad , \quad x \in D_{oh} \tag{3.212}$$

$$V_i^h(x_1,-x_2;\bar{v}^h) = V_i^h(x_1,x_2;\bar{v}^h) \quad , \quad i=0,1, \quad x \in D_{ox}. \tag{3.213}$$

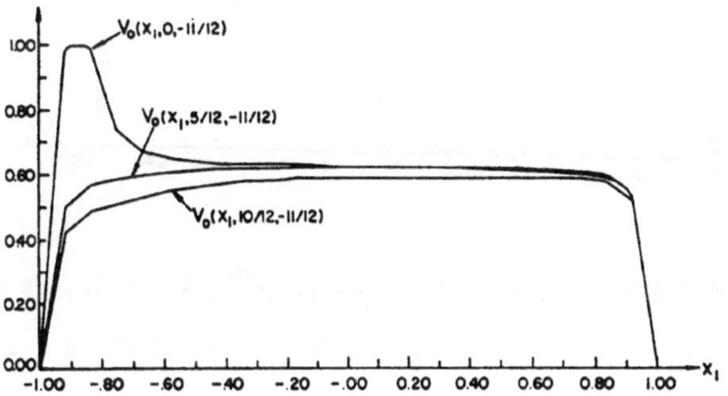

Fig.3.5: The plots of $V_o(x_1,ih,-11/12) = V_o^h(x_1,ih,-11/12;\bar{v}^h)$, i=0,5,10, as functions of x_1, for $u_o = 0.05$, $v_o = 0.05$, $f(x_3) = 0.05$, $q = 1$.

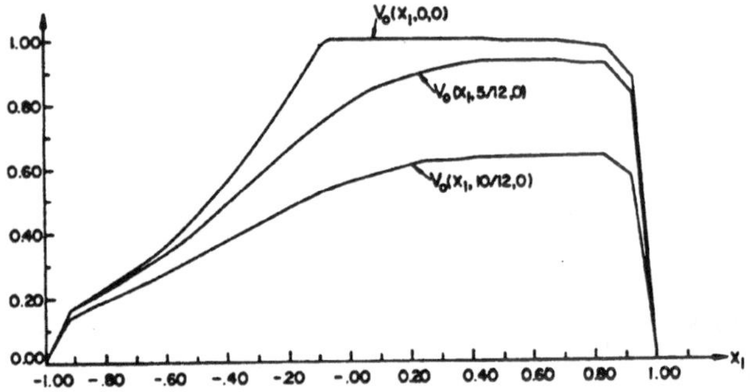

<u>Fig.3.6</u>: The plots of $V_0(x_1,ih,0) = V_0^h(x_1,ih,0;\bar{v}^h)$, i=0,5,10, as func= tions of x_1, for $u_0 = 0.05$, $v_0 = 0.05$, $f(x_3) = 0.05$, q = 1.

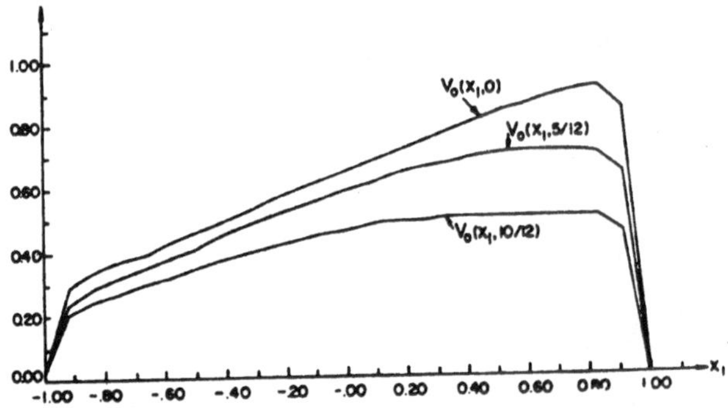

<u>Fig. 3.7</u>: The plots of $V_0(x_1,ih) = V_0^h(x_1,ih;\bar{v}^h)$, i=0,5,10, as functions of x_1, for $u_0 = 0.05$, $v_0 = 0.05$, $f(x_3) = 0.05$, q = 1.

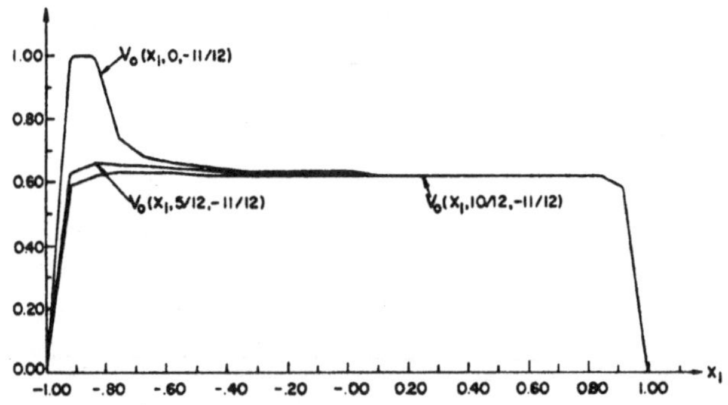

Fig.3.8: The plots of $V_0(x_1,ih,-11/12) = V_0^h(x_1,ih,-11/12;\bar{v}^h)$, i=0,5,10, as functions of x_1, for $u_0 = 0.1$, $v_0 = 0.1$, $f(x_3) = 0.05$, $q = 1$.

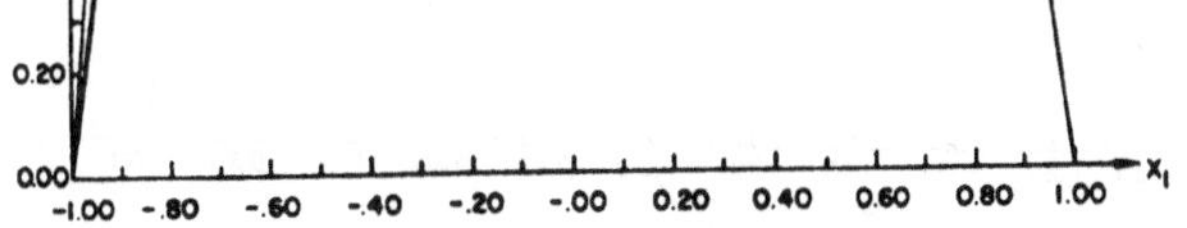

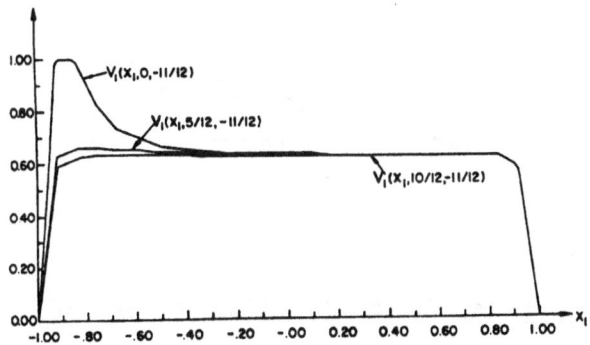

Fig.3.9: The plots of $V_1(x_1,ih,-11/12) = V_1^h(x_1,ih,-11/12;\bar{v}^h)$, i=0,5,10, as functions of x_1, for $u_0 = 0.1$, $v_0 = 0.1$, $f(x_3) = 0.05$, $q = 1$.

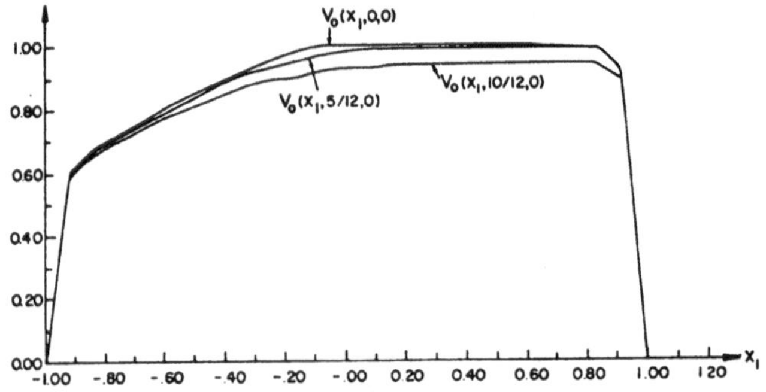

Fig.3.10: The plots of $V_0(x_1,ih,0) = V_0^h(x_1,ih,0;\bar{v}^h)$, i=0,5,10, as functions of x_1, for $u_0 = 0.1$, $v_0 = 0.1$, $f(x_3) = 0.05$, $q = 1$.

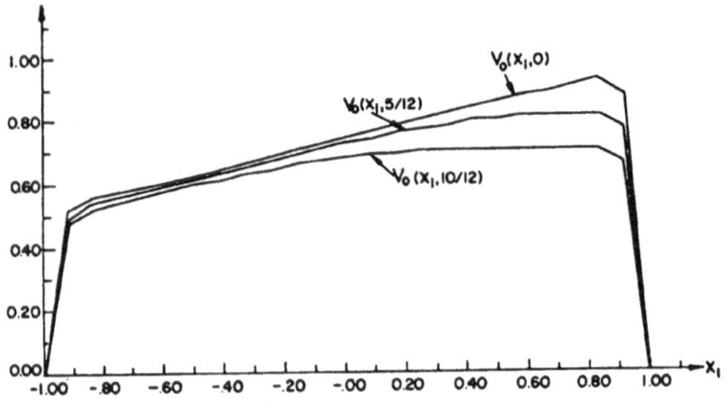

Fig.3.11: The plots of $V_0(x_1,ih) = V_0^h(x_1,ih;\bar{v}^h)$, i=0,5,10, as functions of x_1, for $u_0 = 0.1$, $v_0 = 0.1$, $f(x_3) = 0.05$, $q = 1$.

3.10 EXAMPLE 4 : STRATEGIES USING RANDOM SAMPLING OF THE OBSERVATIONS

The following example is taken from Yavin [103].

3.10.1 Statement of the Problem

Consider the random motion of a point M in the plane, and suppose that the velocity $v = (v_1,v_2)$ of M is perturbed by an $\mathbb{R}^2$-valued Gaussian white noise. It is assumed that ℓ targets are presented in the plane, located at the points $c_1,c_2,\ldots,c_\ell$. The observations available to the point M consist of a sum of measurements, each measurement corresponding to a detection that originated from a target, and all detections arri= ving at random times; plus an $\mathbb{R}^2$-valued Gaussian white noise. Thus, using the observations, the point M cannot associate with the various targets the observed detections which yield the measurements. Using these observations, the point M wishes, by choosing an appropriate velo= city law (strategy) v^*, to steer itself into an ε-neighbourhood of the point c_1, before it leaves an open and bounded domain which contains all the points $c_1,\ldots,c_\ell$.

The problem described above is a simplified model of situations rela= ted to the tracking of a target in a multitarget environment, situations that are described in Bar-Shalom [104], Bar-Shalom and Marcus [105], and the references cited there. However, the methods used here are entirely different from those used in [104] and [105].

We assume here that the points $c_1,\ldots,c_\ell$ are known to the designer of the velocity law (strategy) v, and that the arrival times of the measure= ments are determined by an $\mathbb{R}^\ell$-valued Poisson process. More precisely, the motion of M is given by

$$dx_i = v_i(y)dt + \sigma_i dW_i \quad , \; t > 0 \quad , \quad i=1,2, \quad y = (y_1,y_2), \qquad (3.214)$$

and the observations are given by

$$dy_i = \sum_{j=1}^{\ell} (x_i - c_{ij})dN_j + \gamma_i dB_i, \quad t > 0, \quad i=1,2 \tag{3.215}$$

where $c_j = (c_{1j}, c_{2j})$, $j=1,\ldots,\ell$; σ_i and γ_i, $i=1,2$, are given positive numbers; on a probability space (Ω,F,P), $W = \{W(t) = (W_1(t),W_2(t)), t \geq 0\}$ and $B = \{B(t) = (B_1(t),B_2(t)),\ t \geq 0\}$ are two $\mathbb{R}^2$-valued standard Wiener processes; and $N_1,\ldots,N_\ell$ are mutually independent Poisson processes with parameters $\lambda_1,\ldots,\lambda_\ell$ respectively. If for some j, $1 \leq j \leq \ell$, $\lambda_j = 0$, we say that there is no target at the point c_j. We here use the notations $N = \{N(t) = (N_1(t),\ldots,N_\ell(t)),\ t \geq 0\}$ and $\lambda = (\lambda_1,\ldots,\lambda_\ell)$. It is assumed that W,B and N are mutually independent.

Denote by $\tilde{U}$ the class of all functions $v = v(y)$, $v = (v_1,v_2): \mathbb{R}^2 \to \mathbb{R}^2$ that satisfy: v_i is measurable and $|v_i(y)| \leq v_{0,i}$, for all $y \in \mathbb{R}^2$, i=1,2.

Let $v \in \tilde{U}$ and $\alpha = (x,y) \in \mathbb{R}^4$. Then eqns (3.214)-(3.215) determine a stochastic process $\zeta^v_\alpha = \{\zeta^v_\alpha(t) = (\zeta^v_{\alpha,1}(t),\zeta^v_{\alpha,2}(t),\zeta^v_{\alpha,3}(t),\zeta^v_{\alpha,4}(t)),\ t \geq 0\}$ such that:

(i) on a probability space (Ω,F,P^v_α), ζ^v_α is a weak solution to (3.214)-(3.215) in the sense that P^v_α is the unique solution to the martingale problem for $\mathcal{L}(v)$ (Stroock [86], Mahno [87], and Komatsu [88]), where the operator $\mathcal{L}(v)$ is given by (3.216);

(ii) $\zeta^v_\alpha(0) = \alpha$ P^v_α - almost surely;

(iii) the family $\{P^v_\alpha,\ \alpha \in \mathbb{R}^4\}$ is strong Markov (Stroock [86]);

and

$$\mathcal{L}(v)V(\alpha) \triangleq \sum_{i=1}^{2} v_i(y)\partial V(\alpha)/\partial x_i + (\tfrac{1}{2})\sum_{i=1}^{2}(\sigma_i^2\ \partial^2 V(\alpha)/\partial x_i^2 + \gamma_i^2\ \partial^2 V(\alpha)/\partial y_i^2)$$
$$+ \sum_{k=1}^{\ell} \lambda_k[V(x_1,x_2,y_1 + x_1 - c_{1k},y_2 + x_2 - c_{2k}) - V(\alpha)] \tag{3.216}$$

for any $V \in C_0^\infty(\mathbb{R}^4)$.

$\mathcal{L}(v)$ is the weak infinitesimal operator of the family $\{(\zeta^v_\alpha,P^v_\alpha),\alpha \in \mathbb{R}^4\}$ of strong Markov processes.

We choose here, without loss of generality, $c_1 = (0,0)$. Also, it is assumed here that $d(c_i,c_j) > 2\varepsilon$, $i,j = 1,\dots,\ell$, $i \neq j$ ($d(c_i,c_j)$ denotes the distance between c_i and c_j) and that all the points c_i, $i=2,\dots,\ell$ are within the set $\{x : |x_i| < L,\ i=1,2\}$, where L is a given positive number.

Denote by $D_o(L)$, $K(L)$ and $D(L)$ the following sets in $\mathbb{R}^4$:

$$D_o(L) \triangleq \{\alpha = (x,y) : |x_i| < L \text{ and } |y_i| < L,\ i=1,2\}; \qquad (3.217)$$

$$K(L) \triangleq \{\alpha = (x,y) : x_1^2 + x_2^2 \le \varepsilon^2 \text{ and } |y_i| \le L - \delta,\ i=1,2\}; \qquad (3.218)$$

$$D(L) \triangleq D_o(L) - K(L); \qquad (3.219)$$

where $\varepsilon > 0$ and $0 < \delta << L$ are given numbers. If for some $t \ge 0$, $\zeta_\alpha^v(t) \notin D(L)$, the subsequent motion is disregarded. Define

$$\tau_L(\alpha;v) \triangleq \begin{cases} \inf\{t : \zeta_\alpha^v(t) \notin D(L) \text{ when } \zeta_\alpha^v(0) = \alpha \in D(L)\} \\ 0 \qquad \text{if } \zeta_\alpha^v(0) = \alpha \notin D(L) \\ \infty \qquad \text{if } \zeta_\alpha^v(t) \in D(L) \text{ for all } t \ge 0 \end{cases} \qquad (3.220)$$

and denote

$$U_L \triangleq \{v \in \tilde{U} : \sup_{\alpha \in D(L)} E_\alpha^v\, \tau_L(\alpha;v) < \infty\} \qquad (3.221)$$

$$V_L(\alpha;v) \triangleq P_\alpha^v(\{\zeta_\alpha^v(\tau_L(\alpha;v)) \in K(L)\}), \alpha \in \bar{D}_o(L),\quad v \in U_L. \qquad (3.222)$$

$V_L(\alpha;v)$ is the probability (P_α^v) of the event $\&_L = \{\zeta_\alpha^v$ reaches the set $K(L)$ before it leaves $D_o(L) |$ the strategy $v \in U_L$ is applied and $\zeta_\alpha^v(0) = \alpha\}$. Thus we can write

$$V_L(\alpha;v) \triangleq P_\alpha^v(\&_L) \quad , \alpha \in \bar{D}_o(L) \quad , \quad v \in U_L. \qquad (3.223)$$

Obviously, $P_\alpha^v(\&_L) \le P_\alpha^v(\&_{L_1}) \le \dots \le P_\alpha^v(\&_{L_n})$, $L \le L_1 \le \dots \le L_n < \infty$, $\alpha \in D_o(L)$, $v \in U_L$.

Let $\mathcal{D}_L$ denote the class of all functions $V = V(\alpha)$ such that: V is continuous on the closure $\bar{D}_0(L)$ of $D_0(L)$, and twice continuously differentiable on D(L); for any $v \in U_L$, $\mathcal{L}(v)V \in L_2(D_0(L))$, where $\mathcal{L}(v)$ is given by (3.216).

Given $v \in U_L$. Let $V \in \mathcal{D}_L$ satisfy

$$\mathcal{L}(v)V(\alpha) = 0 \quad , \alpha \in D(L); \tag{3.224}$$

$$V(\alpha) = 1, \alpha \in K(L); \quad V(\alpha) = 0, \alpha \notin D_0(L); \tag{3.225}$$

from which it follows (see, for example, Yavin and Reuter [91]) that

$$V(\alpha) = V_L(\alpha;v) = P_\alpha^v(\{\zeta_\alpha^v(\tau_L(\alpha;v)) \in K(L)\}), \ \alpha \in \bar{D}_0(L). \tag{3.226}$$

Denote

$$U_{oL} \triangleq \{v \in U_L : V_L(\cdot;v) \in \mathcal{D}_L \text{ and satisfies eqns (3.224)-(3.225)}\}, \tag{3.227}$$

$$M_L(v) \triangleq \int_{\bar{D}_0(L)} (1 - V_L(\alpha;v))^2 d\alpha \ , \ v \in U_{oL}. \tag{3.228}$$

In this section the following problem is considered: Find a strategy $v^o \in U_{oL}$ such that

$$M_L(v^o) \leq M_L(v) \text{ for any } v \in U_{oL}. \tag{3.229}$$

The problem dealt with in this section, owing to the form of $\mathcal{L}(v)$ (eq (3.216)), turns out to be a special case of the problem treated in Sections 2.5 and 2.6. Hence, the algorithm for computing weak suboptimal strategies, described in Section 2.7, can be applied to the problem considered in this section. Note that here, the operator $\mathcal{L}^*Q$ (eq (2.79)) assumes the following form

$$\mathcal{L}^*Q(\alpha) \triangleq (\tfrac{1}{2}) \sum_{i=1}^{2} (\sigma_i^2 \, \partial^2 Q(\alpha)/\partial x_i^2 + \gamma_i^2 \, \partial^2 Q(\alpha)/\partial y_i^2)$$

$$+ \sum_{k=1}^{\ell} \lambda_k [Q(x_1, x_2, y_1 - x_1 + c_{1k}, y_2 - x_2 + c_{2k}) - Q(\alpha)] \quad (3.230)$$

$\alpha \in D_0(L)$, for all functions Q such that $\mathcal{L}^*Q \in L_2(D_0(L))$.

3.10.2 Results

The algorithm for computing weak suboptimal strategies, suggested in Section 2.7, has been applied using the following set of parameters: $\sigma_1^2 = \sigma_2^2 = 10^{-6}$, $\gamma_1^2 = \gamma_2^2 = 10^{-4}$, $0.01 \le v_0 \le 0.2$, where $v_{0,1} = v_{0,2} = v_0$; $L = 1, 4/3$; $c_1 = (0,0)$, $c_2 = (2/3,2/3)$, $c_3 = (-2/3,0)$; $\lambda = (\gamma,0,0)$, $0 \le \gamma \le 1$; $\lambda = (0.5,0.5,0)$, $(1,1,0)$, $(0.5,0.5,0.5)$, $(1,1,1)$; $\varepsilon = 0.2$ and $h = 1/6$. In all the cases computed, $v^{(n)}$ converged to $\bar{v}^h$ ($\bar{v}^h$ denotes the limit $\lim_{n \to \infty} v^{(n)}$ on the grid D_{0h} on D_0, with mesh size h along all axes) for $n \le 10$.

Typical extracts from the numerical results are presented in the figures and tables below.

For all the cases where $\lambda = (\gamma,0,0)$, $0 \le \gamma \le 1$; $L = 1, 4/3$ and $0.01 \le v_0 \le 0.2$, it turned out that

$$\bar{v}_i^h(y) = -v_0 \, \mathrm{sign}(y_i) \quad , \quad y \in D_{0y}(L), \quad i=1,2, \qquad (3.231)$$

where sign(0) = 0. The behaviour of $\bar{v}^h$, for some other values of λ, is illustrated in Figs. 3.12-3.17.

The property: $P_\alpha^v(\&_L) \le P_\alpha^v(\&_{L_1})$, $L < L_1$, $\alpha \in D_0(L)$, $v \in U_L$ is illustrated in Table 3.28, where $\alpha^{(1)} = (\pm 1/3,0,0,0)$ or $(0, \pm 1/3,0,0)$ $\alpha^{(2)} = (\pm 1/6, \pm 1/6,0,0)$, $\alpha^{(3)} = (\pm 1/3, \pm 1/3,0,0)$ and $\alpha^{(4)} = (\pm\tfrac{1}{2}, \pm\tfrac{1}{2},0,0)$.

TABLE 3.28: $V_L^h(\alpha^{(i)};\bar{v}^h)$ as functions of v_0 for L = 1, 4/3, λ = (1,0,0) and h = 1/6.

v_0	L=1 $\alpha^{(1)}$	$\alpha^{(2)}$	$\alpha^{(3)}$	$\alpha^{(4)}$	L=4/3 $\alpha^{(1)}$	$\alpha^{(2)}$	$\alpha^{(3)}$	$\alpha^{(4)}$
0.04	.350	.858	.108	.007	.476	.935	.198	.018
0.08	.543	.964	.289	.045	.691	.990	.450	.103
0.12	.662	.988	.430	.102	.803	.997	.607	.209
0.16	.739	.994	.529	.162	.866	.999	.699	.306
0.20	.793	.997	.600	.217	.905	.999	.756	.385

Define the following functionals

$$I(\bar{v}^h) \triangleq \sum_{i,j,k,\ell} V_L^h(ih,jh,kh,\ell h;\bar{v}^h)/(N_h-1)^4 \simeq \int_{D_0} V_L^h(\alpha;\bar{v}^h)d\alpha/\int_{D_0} d\alpha \quad (3.232)$$

$$M_L^h(\bar{v}^h) \triangleq h^4 \sum_{i,j,k,\ell} (1 - V_L^h(ih,jh,kh,\ell h;\, \bar{v}^h))^2 \quad (3.233)$$

$$K(\bar{v}^h) \triangleq \sum_{i,j} V_L^h(ih,jh,0,0;\bar{v}^h)/(N_h-1)^2 \simeq \int_{D_{ox}} V_L^h(x,0;\bar{v}^h)dx/\int_{D_{ox}} dx \quad (3.234)$$

where N_h denotes the number of points along each of the axes in D_{oh}.

Table 3.29 illustrates the dependence of $V_L^h(\cdot;\bar{v}^h)$ on λ in the case where $\lambda = (\gamma,0,0)$.

TABLE 3.29: $I(\bar{v}^h)$, $M_L^h(\bar{v}^h)$ and $K(\bar{v}^h)$ as functions of $\lambda = (\gamma,0,0)$ for $v_{0,1} = v_{0,2} = v_0 = 0.2$, L = 1 and h = 1/8.

γ	$I(\bar{v}^h)$	$M_L^h(\bar{v}^h)$	$K(\bar{v}^h)$
0	.1339	17.5942	.1071
0.05	.1434	17.2457	.4401
0.1	.1499	17.0050	.4507
0.15	.1545	16.8338	.4393
0.2	.1573	16.7207	.4230
0.25	.1587	16.6553	.4056
0.5	.1520	16.7491	.3282
0.75	.1375	17.0965	.2724
1.00	.1236	17.4477	.2325

Fig 3.18 shows the plots of $K(\bar{v}^h)$ as functions of v_0 for three cases: $\lambda = (1,0,0)$, $(1,1,0)$ and $(1,1,1)$.

In order to assess the accuracy of the numerical method, the values of $M_L^h(\bar{v}^h)$ were computed for $h = 1/4, 1/5, 1/6, 1/7$. The results are given in Table 3.30. We denote by $N(h)$ the number of points on the grid on $\bar{D}_0$.

TABLE 3.30: $M_L^h(\bar{v}^h)$ as function of h for: (a) $v_0 = 0.2$, $\lambda = (0.5,0,0)$; (b) $v_0 = 0.5$, $\lambda = (1,0,0)$. In both cases $L = 1$.

h	$M_L^h(\bar{v}^h)$(case (a))	$M_L^h(\bar{v}^h)$(case (b))	N(h)
1/4	23.5909	23.5435	6561
1/5	19.6364	19.5185	14641
1/6	18.4619	18.3551	28561
1/7	17.7018	17.6075	50625

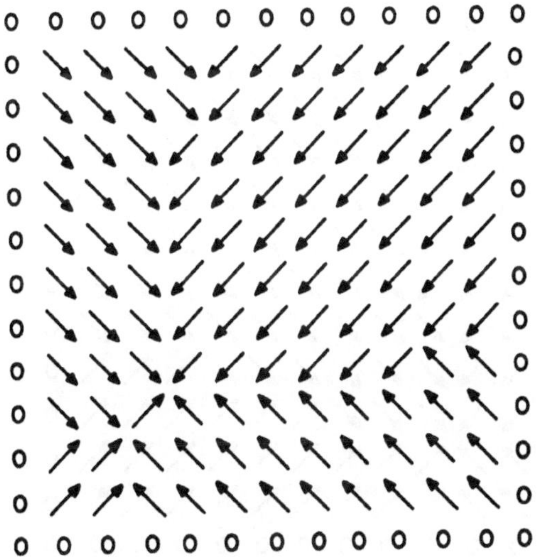

Fig.3.12: The strategy $\bar{v}^h(y) = (\bar{v}_1^h(y), \bar{v}_2^h(y))$ for $y = (k/6, \ell/6)$, $k,\ell = 0, \pm 1, \ldots, \pm 6$, where $v_{0,1} = v_{0,2} = 0.08$, $\lambda = (0.5,0.5,0.0)$ and $L = 1$.

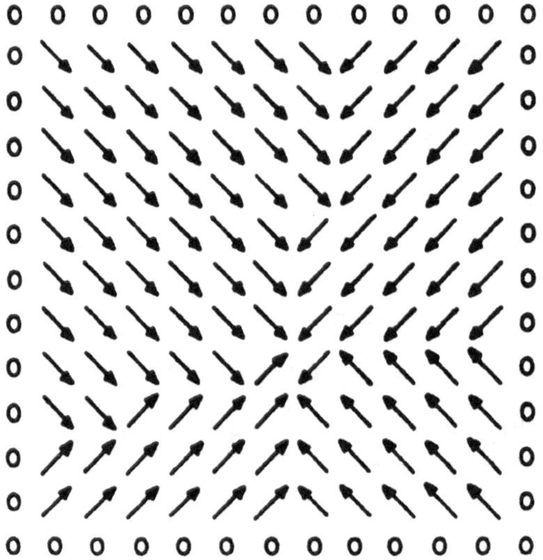

Fig.3.13: The strategy $\bar{v}^h(y) = (\bar{v}_1^h(y), \bar{v}_2^h(y))$ for $y = (k/6, \ell/6)$, $k, \ell = 0, \pm 1, \ldots, \pm 6$ where $v_{o,1} = v_{o,2} = 0.08$, $\lambda = (0.5, 0.5, 0.5)$ and $L = 1$.

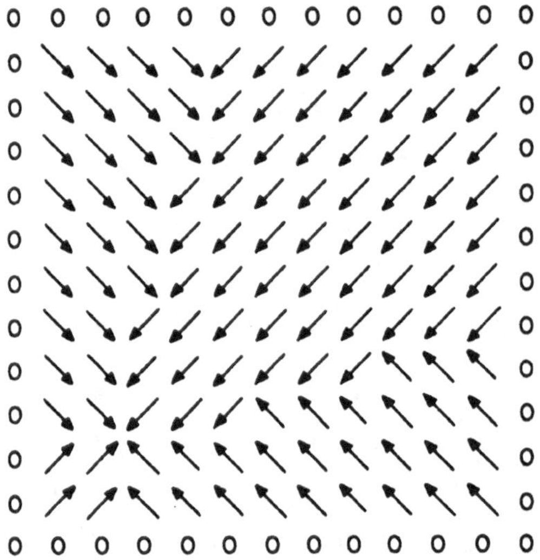

Fig.3.14: The strategy $\bar{v}^h(y) = (\bar{v}_1^h(y), \bar{v}_2^h(y))$ for $y = (k/6, \ell/6)$, $k, \ell = 0, \pm 1, \ldots, \pm 6$, where $v_{o,1} = v_{o,2} = 0.04$, $\lambda = (1,1,0)$ and $L = 1$.

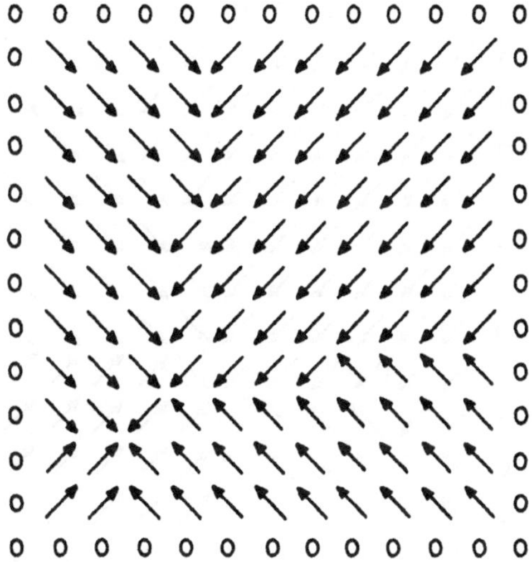

Fig.3.15: The strategy $\bar{v}^h(y) = (\bar{v}_1^h(y), \bar{v}_2^h(y))$ for $y = (k/6, \ell/6)$, $k, \ell = 0, \pm 1, \ldots, \pm 6$, where $v_{o,1} = v_{o,2} = 0.08$, $\lambda = (1,1,0)$ and $L = 1$.

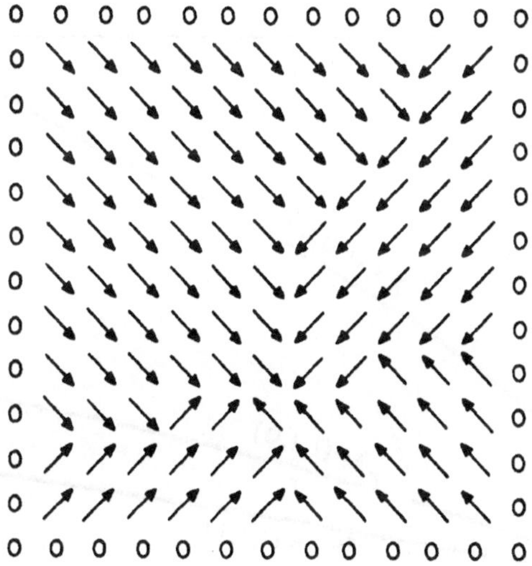

Fig.3.16: The strategy $\bar{v}^h(y) = (\bar{v}_1^h(y), \bar{v}_2^h(y))$ for $y = (k/6, \ell/6)$, $k, \ell = 0, \pm 1, \ldots, \pm 6$, where $v_{o,1} = v_{o,2} = 0.08$, $\lambda = (1,1,1)$ and $L = 1$.

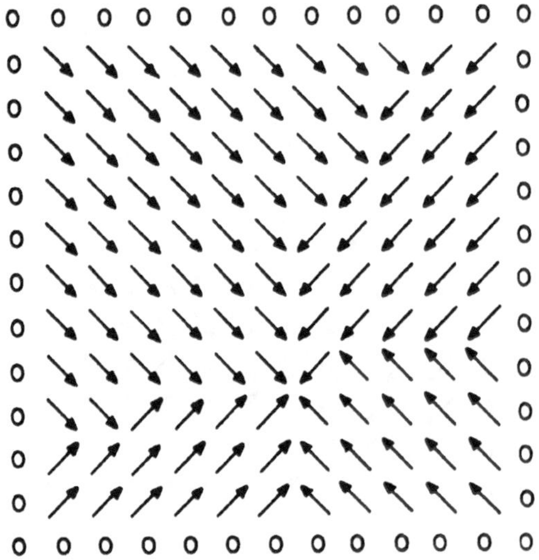

Fig.3.17: The strategy $\bar{v}^h(y) = (\bar{v}_1^h(y), \bar{v}_2^h(y))$ for $y = (k/6, \ell/6)$, $k,\ell = 0, \pm 1, \ldots, \pm 6$, where $v_{0,1} = v_{0,2} = 0.12$, $\lambda = (1,1,1)$ and $L=1$.

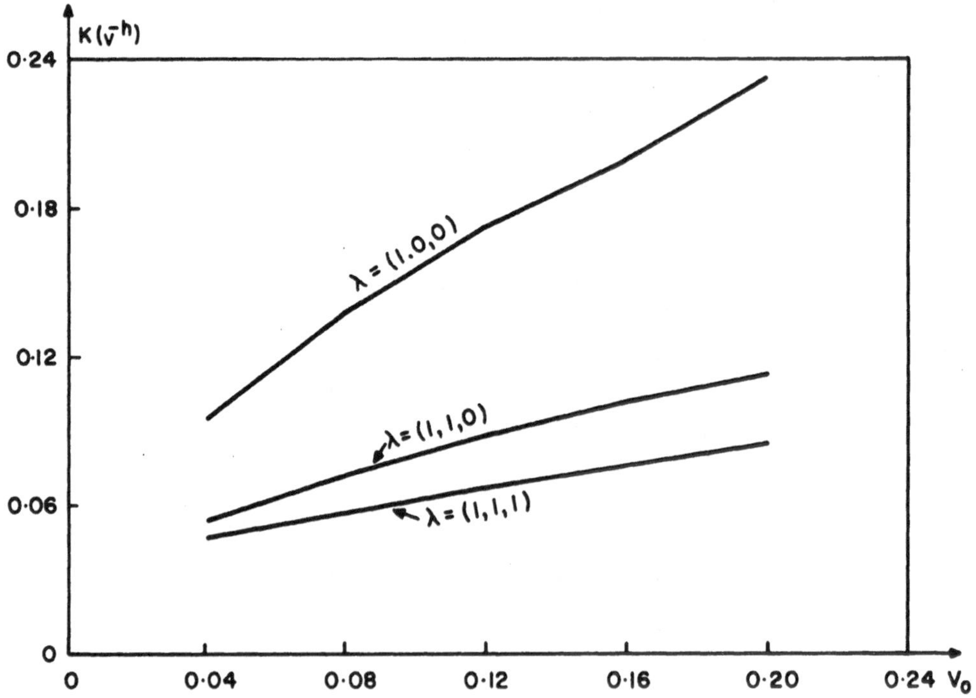

Fig.3.18: $K(\bar{v}^h)$ as function of $v_{0,1} = v_{0,2} = v_0$, where $h = 1/6$ and $L = 1$.

3.10.3 Conclusions

For all the cases where $\lambda = (\gamma,0,0)$, $0 \le \gamma \le 1$; $L = 1$, $4/3$ and $0.01 \le v_o \le 0.2$, it turned out that $\bar{v}^h$ is given by (3.231). Keeping in mind the discussion given in subsection 3.8.3, this result suggests that in the problem where, by using the observations described by (3.215) and in the case where $\lambda = (\gamma,0,0,\ldots,0)$, $\gamma > 0$, the point M has to be steered in such a way that it reaches Δ_T (eq (3.181)), the weak suboptimal stra= tegy $\bar{v}^h$ satisfies a 'separation principle' in the sense described in Subsection 3.8.3.

Denote by $\bar{v}$ a strategy obtained by interpolation of $\bar{v}^h$ such that $\bar{v} \in U_{oL}$.

The results presented in Subsection 3.10.2, and other results as well, indicate that $P_\alpha^{\bar{v}}(\&_L)$ increases when v_o increases. The plots given in Fig.3.18 illustrate the 'jamming' effect of the points $c_2,\ldots,c_\ell$, on the observation available to M.

Table 3.29 illustrates the complicated nature of the information structure given by (3.215). The results given there suggest that $P_\alpha^{\bar{v}}(\&_L)$ is not a monotonic function of $|\lambda|$.

The results obtained here could be improved if we allowed the stra= tegies to depend also on the entire past of $Y_\alpha^v(t) = (\zeta_{\alpha,3}^v(t),\zeta_{\alpha,4}^v(t))$, or part of its past. Unfortunately, this approach is not feasible for implementation owing to computing space restrictions in present-day com= puters.

3.11 PROBABILISTIC INTERPRETATION OF (Q_0,Q_1)

In this section, the probabilistic interpretation of (Q_0,Q_1) as emerges from eqns (3.99), is discussed.

Consider the case where

$$\begin{cases} \partial f_i(x)/\partial x_i = 0 \quad , x \in \mathbb{R}^m \quad , \quad i=1,\dots,m \\ \\ \partial \bar{\sigma}_{ij}(x)/\partial x_j = 0 \quad , \quad x \in \mathbb{R}^m \quad , \quad i,j=1,\dots,m \end{cases} \tag{3.235}$$

as happens to be the case in Example 2 (Section 3.8). Then eqns (3.99) reduce to

$$\begin{cases} -\sum_{i=1}^{m} f_i(x)\partial Q_o(\alpha)/\partial x_i + (\tfrac{1}{2}) \sum_{i,j=1}^{m} [\bar{\sigma}_{ij}(x)\ \partial^2 Q_o(\alpha)/\partial x_i \partial x_j \\ \\ + \bar{\gamma}_{ij}(x)\partial^2 Q_o(\alpha)/\partial y_i \partial y_j] - qQ_o(\alpha) + qQ_1(\alpha) = 1 - V_o(\alpha;v^o) \text{ a.e. in } D_o \\ \\ -\sum_{i=1}^{m} (f_i(x)\partial Q_1(\alpha)/\partial x_i + x_i \partial Q_1(\alpha)/\partial y_i) \\ \\ + (\tfrac{1}{2}) \sum_{i,j=1}^{m} [\bar{\sigma}_{ij}(x)\partial^2 Q_1(\alpha)/\partial x_i \partial x_j + \bar{\gamma}_{ij}(x)\partial^2 Q_1(\alpha)/\partial y_i \partial y_j] - qQ_1(\alpha) \\ \\ + qQ_o(\alpha) = 1 - V_1(\alpha;v^o) \quad , \text{ a.e. in } D_o \end{cases} \tag{3.236}$$

and

$$Q_o(\alpha) = Q_1(\alpha) = 0 \quad , \alpha \notin D_o. \tag{3.237}$$

Consider the following nonlinear stochastic system:

$$\begin{cases} dX = - f(X)dt + \sigma(X)d\tilde{W} \quad , t > 0 \quad , X \in \mathbb{R}^m \\ \\ d\eta = \tilde{\theta}\ Xdt + \gamma(X)d\tilde{B} \quad , t > 0 \quad , \eta \in \mathbb{R}^m \end{cases} \tag{3.238}$$

where, on a probability $(\tilde{\Omega},\tilde{F},\tilde{P})$, $\tilde{W} = \{\tilde{W}(t) = (\tilde{W}_1(t),\dots,\tilde{W}_m(t)),\ t \geq 0\}$ and $\tilde{B} = \{\tilde{B}(t) = (\tilde{B}_1(t),\dots,\tilde{B}_m(t)),\ t \geq 0\}$ are $\mathbb{R}^m$-valued standard Wiener processes and $\tilde{\Theta} = \{\tilde{\theta}(t),\ t \geq 0\}$ is a homogeneous jump Markov process with state space $S = \{0,1\}$ and transition probabilities as given by (3.3). It is assumed that $\tilde{W},\tilde{B}$ and $\tilde{\Theta}$ are mutually independent, and that f, σ and γ satisfy the conditions stated in Section 3.4. Hence, eqns (3.238) have a unique solution $\tilde{\zeta}_\alpha = \{\tilde{\zeta}_\alpha(t) = (X_{\alpha,1}(t),\dots,X_{\alpha,m}(t),\ \eta_{\alpha,1}(t),\dots,\eta_{\alpha,m}(t)),$

$t \geq 0\}$ such that $(\tilde{\zeta}_\alpha, \tilde{\Theta})$ is a strong Markov process and $\tilde{\zeta}_\alpha(0) = \alpha = (x,y) \in \mathbb{R}^{2m}$.

Define

$$\tilde{\tau}_i(\alpha) \triangleq \begin{cases} \inf\{t : (\tilde{\zeta}_\alpha(t),\tilde{\theta}(t)) \in \partial D_o \times S \text{ when } (\tilde{\zeta}_\alpha(0),\tilde{\theta}(0))=(\alpha,i) \in D_o \times S\} \\ 0 \qquad \text{if } \tilde{\zeta}_\alpha(0) = \alpha \notin D_o \text{ and } \tilde{\theta}(0) = i \\ \infty \qquad \text{if } \tilde{\zeta}_\alpha(t) \in D_o \text{ for all } t \geq 0 \text{ and } \tilde{\theta}(0) = i \end{cases} \tag{3.239}$$

$i=0,1$.

Let $\mathcal{D}$ be defined as in Section 3.2. Assume that an element $(Q_o,Q_1) \in \mathcal{D}$ satisfies eqns (3.236) for all $\alpha \in D_o$, and (3.237). Then, by using the techniques leading to (3.18) and Lemma 3.1, it can be shown that

$$Q_i(\alpha) = -\tilde{E}_{\alpha,i} \int_0^{\tilde{\tau}_i(\alpha)} (1 - V_{\tilde{\theta}(t)}(\tilde{\zeta}_\alpha(t);v^o))dt \quad , \alpha \in \bar{D}_o, \ i=0,1, \tag{3.240}$$

where

$$\tilde{E}_{\alpha,i} = \tilde{E}\,[\cdot|(\tilde{\zeta}_\alpha(0),\tilde{\theta}(0)) = (\alpha,i)] \tag{3.241}$$

and $\tilde{E}$ denotes the expectation operator with respect to $\tilde{P}$.

Since

$$V_i(\alpha;v^o) = P_{\alpha,i}(\{\zeta_\alpha^{v^o}(\tau_i(\alpha;v^o)) \in K\}), \quad i=0,1, \ \alpha \in \bar{D}_o \tag{3.242}$$

as follows from the assumptions of Theorem 3.4, eqns (3.240) yield

$$Q_i(\alpha) \leq 0 \quad , \quad i=0,1 \quad , \quad \alpha \in \bar{D}_o. \tag{3.243}$$

The numerical results obtained in Example 2 (Section 3.8) indicate that $Q_i(\alpha;v^{(n)}) < 0$, $\alpha \in D_o$, $i=0,1$, $n=0,1,2,\ldots$.

CHAPTER 4

ESTIMATION AND CONTROL FOR NONLINEAR STOCHASTIC SYSTEMS

4.1 NECESSARY CONDITIONS ON OPTIMAL ESTIMATION AND CONTROL

Let a nonlinear stochastic system be given by

$$dx = [f(x) + F(x)v]dt + \sigma(x)dW + \int_{\mathbb{R}^m} c(x,u)q(dt,du) \tag{4.1}$$

$$t > 0, \quad x \in \mathbb{R}^m \ ,$$

$$dy = h(x)dt + \gamma(x)dB \quad , \ t > 0 \quad , \ y \in \mathbb{R}^p, \tag{4.2}$$

where $X_t = (x_1(t),\ldots,x_m(t))'$, $t \geq 0$, is the state vector and $Y_t = (y_1(t),\ldots,y_p(t))'$, $t \geq 0$, is the output measurement vector. $W = \{W(t) = (W_1(t),\ldots,W_m(t)),\ t \geq 0\}$ and $B = \{B(t) = (B_1(t),\ldots,B_p(t)),\ t \geq 0\}$ are $\mathbb{R}^m$-valued and $\mathbb{R}^p$-valued standard Wiener processes. q is a zero mean Poisson random measure on $[0,\infty) \times \mathbb{R}^m$ as described in Section 2.1. Henceforth it is assumed that W, B and q are mutually independent. (see Section 2.1).

$f : \mathbb{R}^m \to \mathbb{R}^m$, $F : \mathbb{R}^m \to \mathbb{R}^{m\times d}$, $\sigma : \mathbb{R}^m \to \mathbb{R}^{m\times m}$, $c: \mathbb{R}^m \times \mathbb{R}^m \to \mathbb{R}^m$ and $P_J: \mathcal{B}(\mathbb{R}^m) \to [0,1]$ are given functions and a probability measure, respec= tively satisfying the conditions stated in Section 2.2. $v : \mathbb{R}^\ell \to \mathbb{R}^d$, is a feedback strategy. $h : \mathbb{R}^m \to \mathbb{R}^p$ and $\gamma : \mathbb{R}^m \to \mathbb{R}^{p\times p}$ are given functions satisfying: h_i, $i=1,\ldots,p$ and γ_{ij}, $i,j=1,\ldots,p$ are bounded and continu= ously differentiable on $\mathbb{R}^m$.

If $\ell=p$ and $v = v(y)$, then, under proper assumptions on v, v will be a feedback strategy for a stochastic system with partial observation of the kind considered in Chapter 2.

In this section, strategies using $Y^t = \{Y_s,\ 0 \leq s \leq t\}$ are dealt with.

This is done by taking $\ell = m$ and introducing a dynamic state estimator of the form

$$dz = [f(z) + F(z)v(z)]dt + G(z)[dy - h(z)dt], \; t > 0, \; z \in \mathbb{R}^m \tag{4.3}$$

where $G(z) \in \mathbb{R}^{m\times p}$, $z \in \mathbb{R}^m$, has yet to be determined; and using strategies of the form $v = v(z)$ in (4.1).

Henceforth in this section, it is assumed for the sake of simplicity, that: $d=m$, $F_{ij}(x)=\delta_{ij}$, $\sigma_{ij}(x) = \delta_{ij}\sigma_i(x)$, $x \in \mathbb{R}^m$, $i,j=1,\dots,m$; and $\gamma_{ij}(x) = \delta_{ij}\gamma_i(x)$, $x \in \mathbb{R}^m$, $i,j=1,\dots,p$.

By substituting dy from (4.2) into (4.3), and setting $v = v(z)$ in (4.1), the following set of nonlinear stochastic differential equations is obtained:

$$\begin{cases} dx_i = [f_i(x) + v_i(z)]dt + \sigma_i(x)dW_i + \int_{\mathbb{R}^m} c_i(x,u)q(dt,du), \; t > 0, i=1,\dots,m \\ dz_i = [f_i(z) + v_i(z) + \sum_{j=1}^{p} g_{ij}(z)(h_j(x)-h_j(z))]dt + d\tilde{B}_i, \; t > 0, i=1,\dots,m \end{cases} \tag{4.4}$$

where

$$d\tilde{B}_i \triangleq \sum_{j=1}^{p} g_{ij}(z)\gamma_j(x)dB_j \quad , \quad i=1,\dots,m. \tag{4.5}$$

Using the notation $\alpha = (x,z)$, equations (4.4) can be written as

$$d\alpha = F(\alpha,v(z),G(z))dt + \sigma(x,G(z))d\tilde{W} + \int_{\mathbb{R}^m} \tilde{c}(x,u)q(dt,du), \; \alpha \in \mathbb{R}^{2m}, \; t > 0, \tag{4.6}$$

where

$$G(z) \triangleq \begin{pmatrix} g_{11}(z) & \dots & g_{1p}(z) \\ \vdots & & \\ g_{m1}(z) & \dots & g_{mp}(z) \end{pmatrix}, \; z \in \mathbb{R}^m \tag{4.7}$$

$$F_i(\alpha,v(z),G(z)) \triangleq \begin{cases} f_i(x) + v_i(z) & i=1,\dots,m \\ f_{i-m}(z) + v_{i-m}(z) + \sum_{j=1}^{p} g_{i-m,j}(z)(h_j(x)-h_j(z)), & i=m+1,\dots,2m \end{cases} \tag{4.8}$$

$$\tilde{W} = \{\tilde{W}(t) = (W_1(t),\dots,W_m(t),\ B_1(t),\dots,B_m(t)),\ t \geq 0\} \tag{4.9}$$

$$(\sigma(x,G(z))d\tilde{W})_i \triangleq \begin{cases} \sigma_i(x)dW_i & ,\ i=1,\dots,m \\ \sum_{j=1}^{p} g_{i-m,j}(z)\gamma_j(x)dB_j, & i=m+1,\dots 2m \end{cases} \tag{4.10}$$

$$\tilde{c}_i(x,u) \triangleq \begin{cases} c_i(x,u) & ,\ i=1,\dots,m \\ 0 & ,\ i=m+1,\dots,2m \end{cases} \tag{4.11}$$

Let $\tilde{U}$ denote the class of all pairs $(v,G) = \{(v(z),G(z)),\ z \in \mathbb{R}^m\}$ such that:

(a) there is an ℓ_0 for which

$$|F(\alpha,v(z),G(z))|^2 + |\sigma(x,G(z))|^2 \leq \ell_0(1 + |\alpha|^2) \text{ for all } \alpha \in \mathbb{R}^{2m}; \tag{4.12}$$

(b) for any $R > 0$ there is a constant c_R such that when $|\alpha| < R$, $|\alpha'| < R$, $\alpha' = (x',z')$

$$|F(\alpha,v(z),G(z))-F(\alpha',v(z'),G(z'))|^2 + |\sigma(x,G(z))-\sigma(x',G(z'))|^2 \leq c_R|\alpha-\alpha'|^2; \tag{4.13}$$

(c) v_i, $i=1,\dots,m$ are bounded and continuously differentiable on $\mathbb{R}^m$; and g_{ij}, $i=1,\dots,m$, $j=1,\dots,p$ are bounded and twice continuously differentiable on $\mathbb{R}^m$.

Here the following notations are used:

$$|\alpha|^2 = \sum_{i=1}^{2m} \alpha_i^2, \quad |F(\alpha,v,G)|^2 = \sum_{i=1}^{2m} F_i^2(\alpha,v,G), \quad |\sigma(x,G)|^2 = \sum_{i=1}^{2m} \sum_{j=1}^{p} \sigma_{ij}^2(x,G).$$

Under the assumptions on f,σ,c,h and γ it follows (Gihman and Skorohod [1]) that given $\alpha \in \mathbb{R}^{2m}$ and $(v,G) \in \tilde{U}$, equations (4.4) have a unique solution $\zeta_\alpha^{v,G} = \{\zeta_\alpha^{v,G}(t) = (X_\alpha^{v,G}(t), Z_\alpha^{v,G}(t)),\ t \geq 0\}$ with right continuous sample paths and such that $\zeta_\alpha^{v,G}(0) = \alpha$.

Denote by D_0 the following open set in $\mathbb{R}^{2m}$:

$$D_0 \triangleq \{\alpha = (x,z) : |x_i| < a_i \text{ and } |z_i| < a_i, \quad i=1,\ldots,m\} \tag{4.14}$$

where $a_i, i=1,\ldots,m$ are given positive numbers. Define

$$\tau(\alpha;v,G) \triangleq \begin{cases} \inf\{t : \zeta_\alpha^{v,G}(t) \notin D_0 \text{ when } \zeta_\alpha^{v,G}(0) = \alpha \in D_0\} \\ 0 \qquad \text{if } \zeta_\alpha^{v,G}(0) = \alpha \notin D_0 \\ \infty \qquad \text{if } \zeta_\alpha^{v,G}(t) \in D_0 \text{ for all } t \geq 0 \end{cases} \tag{4.15}$$

and

$$U \triangleq \{(v,G) \in \tilde{U} : \sup_{\alpha \in D_0} E_\alpha \tau(\alpha;v,G) < \infty\} \tag{4.16}$$

where $E_\alpha = E[\cdot|\zeta_\alpha^{v,G}(0) = \alpha]$.

In many practical control systems, very large values of the state variable, even though their duration is very short, will cause a serious degradation of the performance of the system, or may correspond to a physical failure. For example, a physical failure of a system can occur the very first time that some or all of the state variables exceed certain fixed bounds, whereas the system operates satisfactorily if the state remains within a certain fixed bounded domain (Crandel and Mark [106]). In this situation the designers will be interested in the dynamical beha= viour of the system only over the time interval $[0,\tau)$.

Let $k(x)$, $x \in \mathbb{R}^m$, be a given continuous function on $\mathbb{R}^m$. We use the notation

$$k(\alpha;v,G) \triangleq k(x) + \lambda_0|x-z| + \sum_{i=1}^{m} \lambda_i v_i^2(z) + \sum_{i=1}^{m} \sum_{j=1}^{p} \lambda_{ij} g_{ij}^2(z), \ \alpha \in D_0, \quad (4.17)$$

where λ_i, $i=0,\ldots,m$ and λ_{ij}, $i=1,\ldots,m$, $j=1,\ldots,p$ are given positive numbers. Define the following functionals

$$V(\alpha;v,G) \triangleq E_\alpha \int_0^{\tau(\alpha;v,G)} k(\zeta_\alpha^{v,G}(t);v,G)dt, \ (v,G) \in U, \ \alpha \in D_0, \quad (4.18)$$

and

$$J(v,G) \triangleq \int_{D_0} V(\alpha;v,G)d\alpha \quad , \quad (v,G) \in U. \quad (4.19)$$

In this section the following estimation and control problem is treated:

Find a pair $(v^*,G^*) \in U$ such that

$$J(v^*,G^*) \leq J(v,G) \quad \text{for any } (v,G) \in U. \quad (4.20)$$

Denote by $\mathcal{D}_0$ the set of all functions $V = V(\alpha)$ such that: V is continuous on the closure $\bar{D}_0$ of D_0, and twice continuously differentiable on D_0; $\partial V/\partial x_i$, $\partial V/\partial z_i$, $\partial^2 V/\partial x_i^2$ and $\partial^2 V/\partial z_i \partial z_j$, $i,j=1,\ldots,m$, are in $L_2(D_0)$.

Define, for $V \in \mathcal{D}_0$ and $(v,G) \in U$

$$\begin{aligned}\mathcal{L}(v,G)(\alpha) \triangleq & \sum_{i=1}^{m} [f_i(x) + v_i(z)]\partial V(\alpha)/\partial x_i \\ & + \sum_{i=1}^{m} [f_i(z) + v_i(z) + \sum_{j=1}^{p} g_{ij}(z)(h_j(x) - h_j(z))]\partial V(\alpha)/\partial z_i \\ & + (\tfrac{1}{2}) \sum_{i=1}^{m} \sigma_i^2(x)\partial^2 V(\alpha)/\partial x_i^2 \qquad (4.21) \\ & + (\tfrac{1}{2}) \sum_{i,j=1}^{m} \sum_{\ell=1}^{p} g_{i\ell}(z)g_{j\ell}(z)\gamma_\ell^2(x)\partial^2 V(\alpha)/\partial z_i \partial z_j \\ & + \rho[\int_{\mathbb{R}^m} V(x + c(x,u),z)P_J(du) - V(\alpha)]\end{aligned}$$

where

$$c(x,u) \triangleq (c_1(x,u),\ldots,c_m(x,u)) \,, \quad x,u \in \mathbb{R}^m. \tag{4.22}$$

$\mathcal{L}(v,G)$ is the weak infinitesimal operator of the family of strong Markov processes $\{(\zeta_\alpha^{v,G},P_\alpha),\ \alpha \in D_o\}$.

Throughout this chapter it is assumed that for any $u \in \mathbb{R}^m$ the mapping

$$\alpha = x + c(x,u) \tag{4.23}$$

maps $\mathbb{R}^m$ one-to-one onto itself and that the inverse mapping

$$x = C(\alpha,u) \tag{4.24}$$

is differentiable. Denote by $\Delta(\alpha,u)$ the Jacobian of the transformation (4.24).

Define, for $(v,G) \in U$ and $\alpha \in D_o$

$$\begin{aligned}\mathcal{L}^*(v,G)Q(\alpha) \triangleq & - \sum_{i=1}^{m} \partial[Q(\alpha)(f_i(x) + v_i(z))]/\partial x_i \\ & - \sum_{i=1}^{m} \partial[Q(\alpha)(f_i(z) + v_i(z) + \sum_{j=1}^{p} g_{ij}(z)(h_j(x)-h_j(z))]/\partial z_i \\ & + (\tfrac{1}{2}) \sum_{i=1}^{m} \partial^2[\sigma_i^2(x)Q(\alpha)]/\partial x_i^2 \\ & + (\tfrac{1}{2}) \sum_{i,j=1}^{m} \partial^2[Q(\alpha) \sum_{\ell=1}^{p} g_{i\ell}(z)g_{j\ell}(z)\gamma_\ell^2(x)]/\partial z_i\partial z_j \\ & + \rho[\int_{\mathbb{R}^m} Q(C(x,u),z)\Delta(x,u)P_J(du) - Q(\alpha)]\end{aligned} \tag{4.25}$$

for any Q such that $\mathcal{L}^*(v,G)Q \in L_2(D_o)$.

By applying the same technique as in the proof of Theorem 2.1 the following theorem is obtained.

Theorem 4.1

Suppose there exists a pair $(v^*,G^*) \in U$ such that

$$J(v^*,G^*) \le J(v,G) \text{ for any } (v,G) \in U. \tag{4.26}$$

Let $(v^\varepsilon,G^\varepsilon) = (v^* + \varepsilon\phi, G^* + \varepsilon\Gamma)$, $(\phi,\Gamma) \in U$, $\varepsilon \in [0,\varepsilon_0]$. Assume:

(i) for each $\varepsilon \in [0,\varepsilon_0]$ there is a function $V^\varepsilon \in D_0$ satisfying

$$\mathcal{L}(v^\varepsilon,G^\varepsilon)V^\varepsilon(\alpha) = -k(\alpha;v^\varepsilon,G^\varepsilon),\ \alpha \in D_0\ ;\ V^\varepsilon(\alpha) = 0,\ \alpha \notin D_0\ ; \tag{4.27}$$

(ii) there is a function Q_0 satisfying

$$\mathcal{L}^*(v^*,G^*)Q_0(\alpha) = -1, \text{ a.e. in } D_0;\quad Q_0(\alpha) = 0,\ \alpha \notin D_0\ ; \tag{4.28}$$

(iii) V^ε, $\partial V^\varepsilon/\partial x_i$, $\partial V^\varepsilon/\partial z_i$, $\partial^2 V^\varepsilon/\partial z_i \partial z_j$, $i,j=1,\ldots,m$ converge weakly (in $L_2(D_0)$) as $\varepsilon \downarrow 0$ to V^0, $\partial V^0/\partial x_i$, $\partial V^0/\partial z_i$, $\partial^2 V^0/\partial z_i \partial z_j$, $i,j=1,\ldots,m$ respectively.

Then:

$$v_i^*(z) = -\Big[2\lambda_i \int_{D_{ox}} Q_0(\alpha)dx\Big]^{-1} \int_{D_{ox}} Q_0(\alpha)[\partial V^0(\alpha)/\partial x_i + \partial V^0(\alpha)/\partial z_i]dx \tag{4.29}$$

$i=1,\ldots,m$, $z \in D_{oz}$

and

$$\left\{\begin{aligned} &2\lambda_{1j}g_{1j}^*(z)\int_{D_{ox}} Q_0(\alpha)dx + \sum_{\ell=1}^m g_{\ell j}^*(z)\int_{D_{ox}} Q_0(\alpha)\gamma_j^2(x)(\partial^2 V^0(\alpha)/\partial z_1\partial z_\ell)dx \\ &\qquad = -\int_{D_{ox}} Q_0(\alpha)(h_j(x)-h_j(z))(\partial V^0(\alpha)/\partial z_1)dx \\ &\quad\vdots \\ &2\lambda_{mj}g_{mj}^*(z)\int_{D_{ox}} Q_0(\alpha)dx + \sum_{\ell=1}^m g_{\ell j}^*(z)\int_{D_{ox}} Q_0(\alpha)\gamma_j^2(x)(\partial^2 V^0(\alpha)/\partial z_m\partial z_\ell)dx \\ &\qquad = -\int_{D_{ox}} Q_0(\alpha)(h_j(x) - h_j(z))(\partial V^0(\alpha)/\partial z_m)dx \end{aligned}\right. \tag{4.30}$$

$j=1,\ldots,p$, $z \in D_{oz}$;

where

$$D_{ox} \triangleq \{x : |x_i| < a_i,\ i=1,\ldots,m\}\ ,\ D_{oz} \triangleq \{z : |z_i| < a_i,\ i=1,\ldots,m\}. \tag{4.14'}$$

We introduce the notation:

$$\begin{cases}\Psi_{ikj}(z;V,Q) \triangleq [\int_{D_{ox}} Q(\alpha)dx]^{-1} \int_{D_{ox}} Q(\alpha)\gamma_j^2(x)(\partial^2V(\alpha)/\partial z_i\partial z_k)dx \\ i,k=1,\dots,m \quad , \quad j=1,\dots,p \quad , \quad z \in D_{oz}\end{cases} \tag{4.31}$$

$$\begin{cases}\Theta_{ij}(z;V,Q) \triangleq [\int_{D_{ox}} Q(\alpha)dx]^{-1} \int_{D_{ox}} Q(\alpha)(h_j(x) - h_j(z))(\partial V(\alpha)/\partial z_i)dx \\ i=1,\dots,m \quad , \quad j=1,\dots,p \quad , \quad z \in D_{oz}.\end{cases} \tag{4.32}$$

Thus, if one assumes that a pair $(v^*,G^*) \in U$ exists, for which equation (4.26) is satisfied, and that all the conditions stated in Theorem 4.1 are satisfied, then, in order to implement this pair, the following system of equations has to be solved:

$$\mathcal{L}(v,G)V(\alpha) = -k(\alpha;v,G) \quad , \quad \alpha \in D_o, \tag{4.33}$$

$$\mathcal{L}^*(v,G)Q(\alpha) = -1, \quad \text{a.e. in } D_o \quad , \tag{4.34}$$

$$V(\alpha) = Q(\alpha) = 0, \ \alpha \notin D_o \tag{4.35}$$

where

$$v_i(z) = -[2\lambda_i \int_{D_{ox}} Q(\alpha)dx]^{-1} \int_{D_{ox}} Q(\alpha)[\partial V(\alpha)/\partial x_i + \partial V(\alpha)/\partial z_i]dx \tag{4.36}$$

$$i=1,\dots,m \quad , \quad z \in D_{oz}$$

and

$$\begin{pmatrix} 2\lambda_{1j} + \Psi_{11j}(z;V,Q) & \Psi_{12j}(z;V,Q) & \dots & \Psi_{1mj}(z;V,Q) \\ \Psi_{21j}(z;V,Q) & 2\lambda_{2j} + \Psi_{22j}(z;V,Q) & \dots & \Psi_{2mj}(z;V,Q) \\ \vdots & \vdots & & \vdots \\ \Psi_{m1j}(z;V,Q) & \Psi_{m2j}(z;V,Q) & & 2\lambda_{mj} + \Psi_{mmj}(z;V,Q) \end{pmatrix} \begin{pmatrix} g_{1j}(z) \\ g_{2j}(z) \\ \vdots \\ g_{mj}(z) \end{pmatrix}$$

$$= \; - \begin{pmatrix} \Theta_{1j}(z;V,Q) \\ \Theta_{2j}(z;V,Q) \\ \vdots \\ \Theta_{mj}(z;V,Q) \end{pmatrix} \quad , \quad j=1,\ldots,p \quad , \; z \in D_{oz} \; . \qquad (4.37)$$

Equations (4.33)-(4.37) are a set of coupled nonlinear partial integro-differential equations. Since these equations constitute necessary conditions on (v^*,G^*), it seems that the problem of the existence and uniqueness of solutions to these equations is crucial to the problem dealt with here. However, owing to the complexity of these equations no efforts are here made to establish such conditions. Instead, a numerical example will be solved for various cases.

The following example is taken from Yavin and Friedman [77].

4.2 A NUMERICAL EXAMPLE

Consider the following random motion of a point M_0 along the x_1-axis in the x_1x_2-plane

$$dx_1 = \sigma dW + \int_{-\infty}^{\infty} c_0 u q(dt,du) \quad , \quad t > 0 \qquad (4.38)$$

where c_0 and σ are given positive numbers, $\{W(t),\ t \geq 0\}$ is a one-dimensional standard Wiener process, and q is a zero mean Poisson random measure on $[0,\infty) \times \mathbb{R}$.

Assume that the location of M_0 is observed from the point $N_0 = (0,H)$ in the plane, and that the angle between the x_2-axis and the section N_0M_0 is measured. By introducing a control law v(z) into the random motion and assuming noisy measurements of the angle, the following equations are obtained

$$\begin{aligned} dx &= v(z)dt + \sigma dW + \int_{-\infty}^{\infty} c_0 u q(dt,du) \quad , \quad (x \triangleq x_1) \\ & \qquad\qquad\qquad\qquad\qquad\qquad\qquad\qquad t > 0 \\ dy &= \arctan(x/H)dt + \gamma dB \end{aligned} \qquad (4.39)$$

where y is the output measurement and γ is a given positive number. $\{B(t), t \geq 0\}$ is a one-dimensional standard Wiener process. It is assumed here that $\{W(t), t \geq 0\}$, $\{B(t), t \geq 0\}$ and q are mutually independent.

The dynamic state estimator is here given by

$$dz = v(z)dt + g(z)[dy - \arctan(z/H)dt], \; t > 0 \tag{4.40}$$

and the set D_0 is here taken as

$$D_0 = \{\alpha : |x| < 1 \text{ and } |z| < 1\} \; , \; \alpha = (x,z). \tag{4.41}$$

On the assumptions that there exists a pair $(v^*,G^*) \in U$ for which equation (4.26) is satisfied, and that all the conditions stated in Theorem 4.1 are satisfied, it follows that in order to compute the values of this pair, the following set of equations has to be solved:

$$\begin{cases} v(z)\partial V(\alpha)/\partial x + [v(z) + g(z)(\arctan(x/H) - \arctan(z/H))]\partial V(\alpha)/\partial z \\ + (\sigma^2/2)\partial^2 V(\alpha)/\partial x^2 + (\gamma^2/2)g^2(z)\partial^2 V(\alpha)/\partial z^2 + \rho \int_{-\infty}^{\infty} V(x+c_0 u,z)P_J(du) \\ - \rho V(\alpha) = -[k(x) + \lambda_0|x-z| + \lambda_1 v^2(z) + \lambda_{11} g^2(z)] \; , \; \alpha \in D_0 \end{cases} \tag{4.42}$$

$$\begin{cases} -v(z)\partial Q(\alpha)/\partial x - \partial[Q(\alpha)(v(z) + g(z)(\arctan(x/H) - \arctan(z/H)))]/\partial z \\ + (\sigma^2/2)\partial^2 Q(\alpha)/\partial x^2 + (\gamma^2/2)\partial^2[Q(\alpha)g^2(z)]/\partial z^2 \\ + \rho \int_{-\infty}^{\infty} Q(x - c_0 u,z)P_J(du) - \rho Q(\alpha) = -1 \quad , \; \alpha \in D_0 \end{cases} \tag{4.43}$$

$$V(\alpha) = Q(\alpha) = 0 \quad , \quad \alpha \in D_0^c \tag{4.44}$$

and

$$\begin{cases} v(z) = -[2\lambda_1 \int_{-1}^{1} Q(\alpha)dx]^{-1} \int_{-1}^{1} Q(\alpha)[\partial V(\alpha)/\partial x + \partial V(\alpha)/\partial z]dx \\ z \in (-1,1) \end{cases} \tag{4.45}$$

$$\begin{cases} g(z) = -\int_{-1}^{1} Q(\alpha)[\arctan(x/H) - \arctan(z/H)](\partial V(\alpha)/\partial z)dx/[2\lambda_{11} \int_{-1}^{1} Q(\alpha)dx \\ \qquad + \gamma^2 \int_{-1}^{1} Q(\alpha)(\partial^2 V(\alpha)/\partial z^2)dx] , \quad z \in (-1,1). \end{cases} \tag{4.46}$$

In this example the following jump distribution is considered:

$$P_J(du) = \begin{cases} du/(2\delta) , & |u| \le \delta \\ 0 , & |u| > \delta \end{cases} \tag{4.47}$$

and $k(x) = -1$, $x \in (-1,1)$.

Let (V^o, Q_o, v^*, g^*) be the unique solution to equations (4.42)-(4.46). In order to evaluate the performance of the system given by equations (4.39)-(4.40) when the pair (v^*, g^*) is being applied, the following problems have been solved:

$$\text{(a)} \quad \mathcal{L}(v^*,g^*)T^o(\alpha) = -1 , \ \alpha \in D_o; \ T^o(\alpha) = 0 , \ \alpha \in D_o^c \tag{4.48}$$

$$\text{(b)} \quad \mathcal{L}(0,0)T(\alpha) = -1 , \ \alpha \in D_o; \ T(\alpha) = 0 , \ \alpha \in D_o^c \tag{4.49}$$

$$\text{(c)} \quad \mathcal{L}(v^*,g^*)K^o(\alpha) = -\lambda_o|x-z| , \ \alpha \in D_o; \ K^o(\alpha) = 0 , \ \alpha \in D_o^c \tag{4.50}$$

$$\text{(d)} \quad \mathcal{L}(0,0)K(\alpha) = -\lambda_o|x-z| , \ \alpha \in D_o; \ K(\alpha) = 0 , \ \alpha \in D_o^c \tag{4.51}$$

$$\text{(e)} \quad \mathcal{L}(0,g^*)T_o(\alpha) = -1 , \ \alpha \in D_o; \ T_o(\alpha) = 0 , \ \alpha \in D_o^c \tag{4.52}$$

$$\text{(f)} \quad \mathcal{L}(0,g^*)K_o(\alpha) = -\lambda_o|x-z| , \ \alpha \in D_o; \ K_o(\alpha) = 0 , \ \alpha \in D_o^c, \tag{4.53}$$

where $\mathcal{L}(v,g)V$ is the left-hand side of equation (4.42). From equations (2.17) it follows that

$$T^o(\alpha) = E_\alpha \ \tau(\alpha;v^*,g^*) \tag{4.54}$$

$$T(\alpha) = E_\alpha \ \tau(\alpha;0,0) \tag{4.55}$$

$$T_o(\alpha) = E_\alpha \ \tau(\alpha;0,g^*) \tag{4.56}$$

$$K^o(\alpha) = E_\alpha \int_0^{\tau(\alpha;v^*,g^*)} \lambda_o|X_\alpha^{v^*,g^*}(t) - Z_\alpha^{v^*,g^*}(t)|dt \tag{4.57}$$

$$K(\alpha) = E_\alpha \int_0^{\tau(\alpha;0,0)} \lambda_o |X_\alpha^{0,0}(t) - Z_\alpha^{0,0}(t)| dt \qquad (4.58)$$

$$K_o(\alpha) = E_\alpha \int_0^{\tau(\alpha;0,g^*)} \lambda_o |X_\alpha^{0,g^*}(t) - Z_\alpha^{0,g^*}(t)| dt \qquad (4.59)$$

where $\tau(\alpha;v,g)$ is defined as in (4.15), and $\{\zeta_\alpha^{v,g}(t) = (X_\alpha^{v,g}(t), Z_\alpha^{v,g}(t)),$ $t \geq 0\}$ is the solution to

$$\begin{cases} dx = v(z)dt + \sigma dW + \int_{-\infty}^{\infty} c_o u q(dt,du) \\ \qquad\qquad\qquad\qquad\qquad\qquad\qquad\qquad t > 0 \\ dz = [v(z) + g(z)(\arctan(x/H) - \arctan(z/H))]dt + g(z)\gamma dB \end{cases} \qquad (4.60)$$

The numbers $\int_{D_0} T^0(\alpha)d\alpha$, $\int_{D_0} T(\alpha)d\alpha$ and $\int_{D_0} T_o(\alpha)d\alpha$ here constitute a measure of the stochastic stability of the system given by (4.60) when the pairs (v^*,g^*), $(0,0)$ and $(0,g^*)$ are applied respectively. Similarly the numbers $\int_{D_0} K^0(\alpha)d\alpha$, $\int_{D_0} K(\alpha)d\alpha$ and $\int_{D_0} K_o(\alpha)d\alpha$ here constitute a measure of the performance of the dynamic state observer given by (4.40) when the pairs (v^*,g^*), $(0,0)$ and $(0,g^*)$ are applied respectively.

Equations (4.42)-(4.46) have been transformed into a set of upwind finite differences equations as described in Section 2.8. The finite dif= ferences equations were iterated until the difference between two consecu= tive iterations did not exceed a given tolerance ε_0. This system of equa= tions led to convergence for $\sigma^2 \geq 0.5$, $\gamma^2 \geq 0$, $\lambda_1 \geq 0.3$ and $\lambda_{11} \geq 0$. Gene= rally the values chosen were $\lambda_o = 1$, $c_o\delta = 0.2$, $\rho = H = 0.5$.

Tables 4.1-4.2 demonstrate the convergence of the numerical procedure for a particular case: $\sigma^2 = \gamma^2 = \lambda_1 = \lambda_{11} = 1$ as $h \downarrow 0$. The tolerance was $\varepsilon_0 = 10^{-5}$. (In Table 4.1, ω and 'it' denote the overrelaxation factor and the number of iterations respectively).

Table 4.1: Convergence of V

h	V(0,0)	V(0,.4)	V(0,.8)	V(.2,0)	V(.6,0)	ω	it
0.2	-.673	-.541	-.270	-.633	-.381	1.4	47
0.1	-.666	-.548	-.293	-.629	-.379	1.6	94
0.05	-.665	-.552	-.304	-.628	-.378	1.8	207

Table 4.2: Convergence of Q

h	Q(0,0)	Q(0,.4)	Q(0,.8)	Q(.2,0)	Q(.6,0)
0.2	1.710	.953	.303	1.639	1.070
0.1	1.749	1.073	.394	1.676	1.081
0.05	1.727	1.144	.430	1.654	1.075

The rate of convergence was always somewhere between $O(h^2)$ and $O(h)$, and the values of V seemed to converge faster than those of Q.

Table 4.3 provides the values of $\int_{D_0} T^o(\alpha)d\alpha$, $\int_{D_0} T(\alpha)d\alpha$, $\int_{D_0} K^o(\alpha)d\alpha$ and $\int_{D_0} K(\alpha)d\alpha$ for the particular cases discussed in the previous tables. It also contains $T^o(0,0)$, $T(0,0)$, $K^o(0,0)$, $K(0,0)$ and demonstrates their convergence as $h \downarrow 0$.

Table 4.3: Convergence of T^o,T,K^o,K

h	$T^o(0,0)$	$T(0,0)$	$K^o(0,0)$	$K(0,0)$	$\int T^o$	$\int T$	$\int K^o$	$\int K$
0.2	.990	.990	.315	.317	2.257	2.352	1.061	1.296
0.1	.993	.992	.325	.327	2.457	2.508	1.207	1.449
0.05	.994	.993	.327	.330	2.561	2.581	1.283	1.521

Table 4.4 contains the values of $\int_{D_0} T^o(\alpha)d\alpha$, $\int_{D_0} T(\alpha)d\alpha$, $\int_{D_0} K^o(\alpha)d\alpha$ and

$\int_{D_0} K(\alpha)d\alpha$ for various cases. The results definitely show that $\int_{D_0} K^o(\alpha)d\alpha < \int_{D_0} K(\alpha)d\alpha$ while the values of $\int_{D_0} T^o(\alpha)d\alpha$ and $\int_{D_0} T(\alpha)d\alpha$ are numerically very close.

Table 4.4: The values of $\int T^o$, $\int T$, $\int K^o$, $\int K$ for: $\sigma^2 = 1$, $c_0\delta = 0.1$, $h = 0.1$, $\varepsilon_0 = 10^{-5}$ and $\omega = 1.5$

γ^2	λ_1	λ_{11}	$\int T^o$	$\int T$	$\int K^o$	$\int K$
.1	.3	.1	2.531	2.520	.999	1.456
.1	.5	.05	2.523	2.520	.973	1.456
.2	.5	.5	2.470	2.520	1.119	1.456
.2	.25	.20	2.484	2.520	1.042	1.456
.1	.5	.5	2.475	2.520	1.115	1.456

The cases included in Table 4.4 were also computed for (0,g*), and the results are given in Table 4.5.

Table 4.5: $\int T^o$, $\int T$, $\int K^o$, $\int K$ for (0,g*) where: $\sigma^2 = 1$, $c_0\delta = 0.1$, $h = 0.1$, $\varepsilon_0 = 10^{-5}$ and $\omega = 1.5$

γ^2	λ_1	λ_{11}	$\int T^o$	$\int T$	$\int K^o$	$\int K$
.1	.3	.1	2.414	2.520	1.020	1.456
.1	.5	.05	2.395	2.520	.971	1.456
.2	.5	.5	2.451	2.520	1.183	1.456
.2	.25	.20	2.415	2.520	1.097	1.456
.1	.5	.5	2.470	2.520	1.182	1.456

Various other cases are included in Table 4.6.

Table 4.6: $\int T^o$, $\int T$, $\int K^o$, $\int K$ for various values of σ^2, γ^2, λ_1 and λ_{11} for: $c_o\delta = 0.2$, $h = 0.2$, $\varepsilon_o = 10^{-5}$

σ^2	γ^2	λ_1	λ_{11}	$\int T^o$	$\int T$	$\int K^o$	$\int K$	ω
.75	.75	.3	.3	2.900	3.126	1.189	1.723	1.3
.5	.5	.3	.3	4.344	4.659	1.649	2.567	1.3
.75	.75	.2	.2	2.874	3.126	1.152	1.723	1.1
.75	.75	.15	.15	2.854	3.126	1.128	1.723	1.1
.75	.75	.30	.15	2.901	3.126	1.179	1.723	1.1
.75	.75	.15	.3	2.845	3.126	1.132	1.723	1.1
.6	.6	.2	.2	3.578	3.895	1.380	2.146	1.1
.4	.4	.3	.3	5.365	5.795	1.965	3.193	0.8

When either σ^2 or λ_1 was chosen too 'small', the algorithm failed to converge. In borderline cases such as $\sigma^2 = \gamma^2 = 0.4$, $\lambda_1 = \lambda_{11} = 0.2$ the numerical procedure led to oscillations in part of the region. These oscilla= tionsdid not vanish if the overrelaxation factor was reduced. When σ^2, λ_1 were reduced even further, the oscillation soon turned into total di= vergence, as could have been expected.

Although it is tempting to try to draw some qualitative conclusions from Tables 4.4-4.6, it seems that the only definite deduction it is possible to make, is that $\int_{D_o} K^o(\alpha)d\alpha < \int_{D_o} K(\alpha)d\alpha$ holds for all the cases under discussion.

4.3 AN ALTERNATIVE APPROACH TO NONLINEAR FILTERING: TIME-CONTINUOUS OBSERVATIONS

Let a nonlinear stochastic system be given by (4.1)-(4.2), where v=0, and assume (for the sake of simplicity) that: $\sigma_{ij}(x) = \delta_{ij}\sigma_i(x)$, $x \in \mathbb{R}^m$, $i,j=1,\ldots,m$; and $\gamma_{ij}(x) = \delta_{ij}\gamma_i(x)$, $x \in \mathbb{R}^m$, $i,j=1,\ldots,p$.

Thus the following system is considered

$$dx_i = f_i(x)dt + \sigma_i(x)dW_i + \int_{\mathbb{R}^m} c_i(x,u)q(dt,du)$$

$$t > 0 \quad , \quad x \in \mathbb{R}^m \quad , \quad i=1,\dots,m \tag{4.61}$$

$$dy_i = h_i(x)dt + \gamma_i(x)dB_i, \quad t > 0, \quad i=1,\dots,p, \tag{4.62}$$

where $x(t) \triangleq \text{col.}(x_1(t),\dots,x_m(t))$, $t \geq 0$, is the state vector and $y(t) \triangleq \text{col.}(y_1(t),\dots,y_p(t))$, $t \geq 0$, is the output measurement vector.

For the system given by (4.61)-(4.62), a dynamic state estimator is chosen having the following form:

$$dz_i = f_i(z)dt + \sum_{j=1}^{p} g_{ij}(z)(dy_j - h_j(z)dt), \; t > 0, \; i=1,\dots,m \tag{4.63}$$

where the functions $g_{ij}(z)$, $z \in \mathbb{R}^m$, $i=1,\dots,m$, $j=1,\dots,p$, are yet to be determined. The matrix G

$$G(z) \triangleq \begin{pmatrix} g_{11}(z) & \dots & g_{1p}(z) \\ \cdot & & \\ \cdot & & \\ \cdot & & \\ g_{m1}(z) & \dots & g_{mp}(z) \end{pmatrix} \tag{4.64}$$

will here be called the *gain matrix* of the observer (4.63).

By substituting dy from (4.62) into (4.63), the following set of stochastic differential equations is obtained:

$$\begin{cases} dx_i = f_i(x)dt + \sigma_i(x)dW_i + \int_{\mathbb{R}^m} c_i(x,u)q(dt,du), \; t > 0, \; i=1,\dots,m \\ dz_i = [f_i(z) + \sum_{j=1}^{p} g_{ij}(z)(h_j(x) - h_j(z))]dt + d\tilde{B}_i, \; t > 0, \; i=1,\dots,m \end{cases} \tag{4.65}$$

where

$$d\tilde{B}_i \triangleq \sum_{j=1}^{p} g_{ij}(z)\gamma_j(x)dB_j \quad , \quad i=1,\dots,m. \tag{4.66}$$

Let

$$F_i(\alpha,G(z)) \stackrel{\Delta}{=} \begin{cases} f_i(x) & i=1,\dots,m \\ f_{i-m}(z) + \sum_{j=1}^{p} g_{i-m,j}(z)(h_j(x) - h_j(z)) & i=m+1,\dots,2m \end{cases} \tag{4.67}$$

$\alpha = (x,z)$, and let $\tilde{W}$, $\sigma(x,G(z))$ and $\tilde{c}$ be defined by (4.9),(4.10) and (4.11) respectively.

It is assumed here that f_i, $i=1,\dots,m$ are bounded and continuously differentiable on $\mathbb{R}^m$; that h_j, $j=1,\dots,p$ are bounded and continuous on $\mathbb{R}^m$; that σ_i, $i=1,\dots,m$ and γ_j, $j=1,\dots,p$ are bounded and twice continuously differentiable on $\mathbb{R}^m$; and that c_i, $i=1,\dots,m$ are bounded on $\mathbb{R}^{2m}$, and, for any $u \in \mathbb{R}^m$, are continuously differentiable with respect to x on $\mathbb{R}^m$.

Let $\tilde{P}_J$ be a probability measure on $\mathcal{B}(\mathbb{R}^m)$. Define, for $A \in \mathcal{B}(\mathbb{R}^{2m})$

$$M(\alpha,A) \stackrel{\Delta}{=} \tilde{P}_J(u : \tilde{c}(x,u) \in A),\ \alpha = (x,z) \in \mathbb{R}^{2m}, \tag{4.68}$$

(where $\tilde{c}(x,u) = (c_1(x,u),\dots,c_m(x,u),0,\dots,0) \in \mathbb{R}^{2m}$, see (4.11)).

Assume that

(a) for any $A = A_1 \times A_2 \in \mathcal{B}(\mathbb{R}^m) \times \mathcal{B}(\mathbb{R}^m)$

$$M(\alpha,A) = \begin{cases} P_J(u : c(x,u) \in A_1) & \text{if } A_2 \text{ contains the origin} \\ 0 & \text{otherwise,} \end{cases} \tag{4.69}$$

(where P_J is introduced in Section 2.1);

(b) $\int_A (\alpha'/(1 + |\alpha'|^2))M(\alpha,d\alpha')$ and $\int_{\mathbb{R}^{2m}} (|\alpha'|^2/(1 + |\alpha'|^2))\phi(\alpha')M(\alpha,d\alpha')$ are bounded and continuous on $\mathbb{R}^{2m}$ for any $A \in \mathcal{B}(\mathbb{R}^{2m})$ and all $\phi \in C_b(\mathbb{R}^{2m})$ ($C_b(\mathbb{R}^{2m})$ is the set of bounded continuous functions $f : \mathbb{R}^{2m} \to \mathbb{R}$.)

Note that (4.68) and (4.69) imply that

$$\int_{\mathbb{R}^m} [V(\alpha + \tilde{c}(x,u))-V(\alpha) - \sum_{i=1}^{2m} \tilde{c}_i(x,u)\partial V(\alpha)/\partial\alpha_i]\tilde{P}_J(du)$$

$$= \int_{\mathbb{R}^{2m}} [V(\alpha + \alpha')-V(\alpha) - \sum_{i=1}^{m} \alpha_i'\partial V(\alpha)/\partial x_i]M(\alpha,d\alpha') \tag{4.70}$$

$$= \int_{\mathbb{R}^m} [V(x + c(x,u),z) - V(\alpha) - \sum_{i=1}^{m} c_i(x,u)\partial V(\alpha)/\partial x_i]P_J(du),$$

for any $V \in C^1(\mathbb{R}^{2m})$.

Let $\tilde{U}$ denote the class of all gain matrices $G = \{G(z),\ z \in \mathbb{R}^m\}$ such that:

(i) g_{ij}, $i=1,\ldots,m$, $j=1,\ldots,p$ are measureable and $|g_{ij}(z)| = g_{ij}^o$ a.e. in $\mathbb{R}^m$, where g_{ij}^o, $i=1,\ldots,m$, $j=1,\ldots,p$ are given non-negative num= bers;

(ii) $\sigma(x,G(z))\sigma'(x,G(z)) = \sigma(x,G^o)\sigma'(x,G^o)$ for all $(x,z) \in \mathbb{R}^m \times \mathbb{R}^m$, where $(G^o)_{ij} = g_{ij}^o$, $i=1,\ldots,m$, $j=1,\ldots,p$;

(iii) the matrix $\sigma(x,G^o)\sigma'(x,G^o)$ has a symmetric positive-definite square root $\bar{\sigma}(x,G^o)$;

(iv) $\sigma(x,G^o)\sigma'(x,G^o)$ is continuous in x; there exist constants λ_1 and λ_2, $0 < \lambda_1 \le \lambda_2 < \infty$ such that

$$\lambda_1|\alpha|^2 \le (\alpha,\sigma(x',G^o)\sigma'(x',G^o)\alpha) \le \lambda_2|\alpha|^2$$

for any $x' \in \mathbb{R}^m$, and $\alpha = (x,z) \in \mathbb{R}^m \times \mathbb{R}^m$.

Under the assumptions on $f_i,\sigma_i,g_{ij},h_j,\gamma_j,i=1,\ldots,m$ and $j=1,\ldots,p$, it follows (Stroock [86]) that for a given $G \in \tilde{U}$ eqns (4.65) have a unique weak solution $(\zeta_\alpha^G,P_\alpha^G)$, such that $\zeta_\alpha^G(0) = \alpha$, which is a strong Markov pro= cess.

Let the set D_o be given by (4.14). Define, for $G \in \tilde{U}$

$$\tau_0(\alpha;G) \triangleq \begin{cases} \inf\{t : \zeta_\alpha^G(t) \notin D_0 \text{ when } \zeta_\alpha^G(0) = \alpha \in D_0\} & \\ 0 & \text{if } \zeta_\alpha^G(0) = \alpha \notin D_0 \\ \infty & \text{if } \zeta_\alpha^G(t) \in D_0 \text{ for all } t \geq 0 \end{cases} \tag{4.71}$$

$$U \triangleq \{G \in \tilde{U} : \sup_{\alpha \in D_0} E_\alpha^G \, \tau_0(\alpha;G) < \infty\} , \tag{4.72}$$

and for a given positive number ε, held fixed, set

$$V(\alpha;G) \triangleq E_\alpha^G \, \Lambda \, \{t : 0 \leq t < \tau_0(\alpha;G), \; |\chi_\alpha^G(t) - \eta_\alpha^G(t)| \leq \varepsilon\} \; , \; G \in U \tag{4.73}$$

where $\zeta_\alpha^G(t) = (\chi_{\alpha,1}^G(t),\ldots,\chi_{\alpha,m}^G(t),\eta_{\alpha,1}^G(t),\ldots,\eta_{\alpha,m}^G(t))$; E_α^G denotes the expectation operation with respect to P_α^G; and Λ is the Lebesgue measure on the real line.

Let T be a given positive number such that

$$\sup_{G \in U} E_\alpha^G \, \tau_0(\alpha;G) < T \text{ for all } \alpha \in D_0 \tag{4.74}$$

(it is assumed here that $T < \infty$).

In this section the following estimation problem is treated: Find a gain matrix $G^* \in U$ such that

$$\ell(G^*) \leq \ell(G) \text{ for any } G \in U \tag{4.75}$$

where

$$\ell(G) \triangleq \int_{D_0} (T - V(\alpha;G))^2 d\alpha, \quad G \in U . \tag{4.76}$$

A gain matrix $G^* \in U$ for which (4.75) is satisfied will here be called a *weak optimal gain matrix.*

Let $\mathcal{D}_0$ denote the class of functions $V : \mathbb{R}^{2m} \to \mathbb{R}$ such that: V is bounded and continuous on the closure $\bar{D}_0$ of D_0; twice continuously dif= ferentiable on D_0; for any $G \in U$, $\mathcal{L}(G)V \in L_2(D_0)$, where

$$\mathcal{L}(G)V(\alpha) \triangleq \sum_{i=1}^{m} \{f_i(x)\partial V(\alpha)/\partial x_i + [f_i(z) + \sum_{j=1}^{p} g_{ij}(z)(h_j(x)-h_j(z))]\partial V(\alpha)/\partial z_i\}$$

$$+ (\tfrac{1}{2}) \sum_{i,j=1}^{2m} (\sigma(x,G^o)\sigma'(x,G^o))_{ij}\ \partial^2 V(\alpha)/\partial\alpha_i\ \partial\alpha_j \tag{4.77}$$

$$+ \rho \int_{\mathbb{R}^m} [V(\alpha + \tilde{c}(x,u)) - V(\alpha) - \sum_{i=1}^{m} c_i(x,u)\partial V(\alpha)/\partial x_i]P_J(du)$$

and $x = (\alpha_1,\ldots,\alpha_m)$, $z = (\alpha_{m+1},\ldots,\alpha_{2m})$.

The operator $\mathcal{L}(G)$ is the weak infinitesimal operator of the family $\{(\zeta_\alpha^G, P_\alpha^G),\ \alpha \in D_o\}$ of strong Markov processes.

Denote by A the following set in $\mathbb{R}^{2m}$:

$$A \triangleq \{\alpha = (x,z) : \alpha \in D_o \text{ and } |x-z| \le \varepsilon\} \tag{4.78}$$

and let I_A denote the indicator function of A. Then

$$\Lambda\{t : 0 \le t < \tau_o(\alpha;G), |x_\alpha^G(t)-\eta_\alpha^G(t)| \le \varepsilon\} = \int_0^{\tau_o(\alpha;G)} I_A(\zeta_\alpha^G(t))dt. \tag{4.79}$$

Also, it can be shown that:

Given $G \in U$. Let $V \in \mathcal{D}_o$ satisfy

$$\mathcal{L}(G)V(\alpha) = -I_A(\alpha)\ ,\ \alpha \in D_o;\quad V(\alpha) = 0\ ,\ \alpha \notin D_o\ ; \tag{4.80}$$

then

$$V(\alpha) = V(\alpha;G) = E_\alpha^G\ \Lambda\{t : 0 \le t < \tau_o(\alpha;G),\ |x_\alpha^G(t)-\eta_\alpha^G(t)| \le \varepsilon\}\ . \tag{4.81}$$

In the same manner as in Section 2.6, we define the set U_o,

$$U_o \triangleq \{G \in U : V(\cdot;G) \in \mathcal{D}_o \text{ and satisfies eqns (4.80)}\}\ , \tag{4.82}$$

and the operator $\mathcal{L}^*$

$$\mathcal{L}^*Q(\alpha) \overset{\Delta}{=} - \sum_{i=1}^{m} \{\partial[f_i(x)Q(\alpha)]/\partial x_i + \partial[f_i(z)Q(\alpha)]/\partial z_i\}$$

$$+ (\tfrac{1}{2}) \sum_{i,j=1}^{2m} \partial^2[(\sigma(x,G^o)\sigma'(x,G^o))_{ij}Q(\alpha)]/\partial\alpha_i\, \partial\alpha_j \tag{4.83}$$

$$+ \rho \int_{\mathbb{R}^m} [Q(C(x,u),z)\Delta(x,u) - Q(\alpha) + \sum_{i=1}^{m} \partial[c_i(x,u)Q(\alpha)]/\partial x_i]P_J(du)$$

for all Q such that $\mathcal{L}^*Q \in L_2(D_o)$, where C and Δ are described in Section 4.1.

Applying the same procedure as in the proof of Theorem 2.3, the fol= lowing theorem is obtained.

Theorem 4.2

Suppose $V_o \in \mathcal{D}_o$, $G^* \in U_o$, and Q_o satisfy

$$\mathcal{L}(G^*)V_o(\alpha) = - I_A(\alpha),\ \alpha \in D_o\ ;\ V_o(\alpha) = 0\ ,\ \alpha \notin D_o \tag{4.84}$$

$$\mathcal{L}^*Q_o(\alpha) = T - V_o(\alpha),\ \text{a.e. in } D_o\ ;\ Q_o(\alpha) = 0\ ,\ \alpha \notin D_o \tag{4.85}$$

where $G^* \in U_o$ is determined by

$$G^* = \arg\sup_{G \in U_o} \{- \sum_{i=1}^{m} \sum_{j=1}^{p} \int_{D_{oz}} g_{ij}(z) \int_{D_{ox}} Q_o(\alpha)[h_j(x)-h_j(z)](\partial V(\alpha;G)/\partial z_i)dxdz\} \tag{4.86}$$

and D_{ox} and D_{oz} are given by (4.14') (Theorem 4.1); then

$$\ell(G^*) \le \ell(G) \text{ for any } G \in U_o. \tag{4.87}$$

We are interested here, as in Section 2.6, in the minimization of $\ell(G)$ on U_o only, $U_o \subset U$.

In the same manner as in Chapter 2, an algorithm for computing weak suboptimal gain matrices is suggested below.

1. Given $G^{(n)} \in U_0$

2. Compute $V(\cdot;G^{(n)})$ by solving numerically the following problem:

$$\mathcal{L}(G^{(n)})V(\alpha) = - I_A(\alpha), \quad \alpha \in D_0 \; ; \; V(\alpha) = 0 \quad , \alpha \notin D_0.$$

3. Calculate $\ell(G^{(n)})$.

4. Compute $Q(\cdot;G^{(n)})$ by solving numerically the following problem:

$$\mathcal{L}^*Q(\alpha) = T - V(\alpha;G^{(n)}) \; , \; \alpha \in D_0 \; ; \; Q(\alpha) = 0 \; , \; \alpha \notin D_0.$$

5. Compute $G^{(n+1)}$ by

$$G^{(n+1)} = \arg \sup_{G \in U_0} \{- \sum_{i=1}^{m} \sum_{j=1}^{p} \int_{D_{oz}} g_{ij}(z) \int_{D_{ox}} Q(\alpha;G^{(n)})[h_j(x)-h_j(z)]$$

$$(\partial V(\alpha;G^{(n)})/\partial z_i)dxdz\}$$

6. If $G^{(n+1)} \neq G^{(n)}$; then $n+1 \to n$, and go to 2. Otherwise: stop.

The computations are continued until for some $n \geq 0$ either $G^{(n+1)}=G^{(n)}$ or $\ell(G^{(n+1)})=\ell(G^{(n)})$.

Given $h > 0$, then, whenever the sequence $\{\ell(G^{(n)})\}$ converges (when computations are carried on a grid D_{oh} on D_0) and $G^{(n)} \to \bar{G}^h$, we denote the limit solution by $(V(\cdot;\bar{G}^h), Q(\cdot;\bar{G}^h),\bar{G}^h)$.

Henceforward in this chapter, steps 2 and 4 of the algorithm for com= puting weak suboptimal gain matrices are implemented by using a finite-differences procedure similar to that described in Section 2.8.

An important factor in nonlinear filtering problems is the structure of the measurement process given by (4.62). An interesting question is the following: assume that the designer can choose a measurement policy, i.e. he can choose the function $h(x)$. If several measurement policies, say $h^{(1)}(x),\dots,h^{(k)}(x)$, are available for choice, which of them should the designer choose? The computation, off-line, of the functions $V_i(\alpha;\bar{G})$,

$i=1,\ldots,k$, where $V_i(\alpha;\bar{G})$ is determined by the algorithm for computing weak suboptimal gain matrices, in which $h = h^{(i)}$, and the comparison of the values of the functions, might help the designer to make his choice.

In order to evaluate the performance of the dynamic state estimator

$$dz_i = f_i(z)dt + \sum_{j=1}^{p} \bar{g}_{ij}(z)(dy_j - h_j(z)dt),\ t > 0,\ i=1,\ldots,m \qquad (4.63')$$

where $\bar{G} \in U_0$ is a weak suboptimal gain matrix, the following problem has been solved:

$$\mathcal{L}(\bar{G})T(\alpha) = -1,\ \alpha \in D_0\ ;\ T(\alpha) = 0\ ,\ \alpha \in D_0^c \qquad (4.88)$$

from which it follows that

$$T(\alpha) = E_\alpha^{\bar{G}}\ \tau_0(\alpha;\bar{G})\ ,\ \alpha \in D_0 \qquad (4.89)$$

The functions $T(\alpha)$ here constitute a measure of the stochastic stability of the system given by (4.65) with $G = \bar{G}$.

4.4 NUMERICAL EXAMPLES

The following examples are taken from Yavin [78].

4.4.1 Example 1

Consider the one-dimensional system

$$dx = -\ v_0 dt + \sigma dW\ ,\ t > 0\ ,\ x \in \mathbb{R} \qquad (4.90)$$

with the measurement process $\{y_t,\ t \geq 0\}$ determined by

$$dy = h(x)dt + \gamma dB\ ,\ t > 0,\ \ y \in \mathbb{R}\ , \qquad (4.91)$$

where v_0, σ and γ are given positive numbers, and $\{W(t),\ t \geq 0\}$ and $\{B(t),\ t \geq 0\}$ are two independent one-dimensional Wiener processes. h is a given continuous function. In this case the dynamic state estimator is given by

$$dz = -v_0 dt + g(z)(dy - h(z)dt) \;,\; t > 0 \tag{4.92}$$

and the set D_0 is here chosen to be

$$D_0 \triangleq \{\alpha = (x,z) : |x| < 1 \text{ and } |z| < 1\} . \tag{4.93}$$

Thus equations (4.65) yield

$$\begin{cases} dx = -v_0 dt + \sigma dW \\ dz = [-v_0 + g(z)(h(x) - h(z))]dt + g(z)\gamma dB \;, \end{cases} \tag{4.94}$$

and steps 2,4 and 5 of the algorithm for computing weak suboptimal gain matrices here reduce to

$$-v_0\, \partial V(\alpha)/\partial x + [-v_0 + g^{(n)}(z)(h(x)-h(z))]\partial V(\alpha)/\partial z$$

$$+ (\tfrac{1}{2})[\sigma^2\, \partial^2 V(\alpha)/\partial x^2 + g_0^2\, \gamma^2\, \partial^2 V(\alpha)/\partial z^2] = -I_A(\alpha),\; \alpha \in D_0; \tag{4.95}$$

$$V(\alpha) = 0 \;,\; \alpha \notin D_0$$

$$v_0 \partial Q(\alpha)/\partial x + v_0\, \partial Q(\alpha)/\partial z + (\tfrac{1}{2})[\sigma^2\, \partial^2 Q(\alpha)/\partial x^2 + g_0^2\, \gamma^2\, \partial^2 Q(\alpha)/\partial z^2]$$

$$= T - V(\alpha;g^{(n)}) \;,\; \alpha \in D_0 \;;\; Q(\alpha) = 0 \;,\; \alpha \notin D_0 \tag{4.96}$$

(where $V(\cdot;g^{(n)})$ and $Q(\cdot;g^{(n)})$ denote the solutions to (4.95) and (4.96) respectively)

and

$$g^{(n+1)}(z) = -g_0 \operatorname{sign}\left(\int_{-1}^{1} Q(\alpha;g^{(n)})(h(x)-h(z))(\partial V(\alpha;g^{(n)})/\partial z)dx\right), z \in (-1,1);$$

$$n=0,1,\ldots \;; \tag{4.97}$$

respectively, where $G^0 = g_0$ (see the definition of the set $\tilde{U}$, section 4.3).

Given $h > 0$, then, whenever the sequence $\{\ell(g^{(n)})\}$ converges and $g^{(n)} \to \bar{g}^h$, we denote the limit solution by $(V(\cdot;\bar{g}^h), Q(\cdot;\bar{g}^h), \bar{g}^h)$.

The functions $V(\alpha;\bar{g}^h)$ and $T(\alpha) = E_{\alpha}^{\bar{g}^h} \tau_0(\alpha;\bar{g}^h)$ were computed for the following systems:

4.4.1.a. $dx = -v_0 dt + \sigma dW$, $dy = m_0 x dt + \gamma dB$, $t > 0$ (4.98)

4.4.1.b. $dx = -v_0 dt + \sigma dW$, $dy = m_0 x^3 dt + \gamma dB$, $t > 0$ (4.99)

4.4.1.c. $dx = -v_0 dt + \sigma dW$, $dy = m_0 x^5 dt + \gamma dB$, $t > 0$ (4.100)

4.4.1.d. $dx = -v_0 dt + \sigma dW$, $dy = m_0 x^9 dt + \gamma dB$, $t > 0$ (4.101)

4.4.1.e. $dx = -v_0 dt + \sigma dW$, $dy = m_0 \arctan(x) dt + \gamma dB$, $t > 0$ (4.102)

where $v_0 = 0.2$, $m_0 = 10$, $\sigma^2 = 0.04$, $\gamma^2 = 0.05$ (except the case 4.4.1.e where $\gamma^2 = 0.01$), $g_0 = 5$, $T = 20$ and $h = 0.025$. The set A was taken as

$$A \triangleq \{\alpha = (x,z) : \alpha \in D_0 \text{ and } |x-z| \le 0.01\}. \quad (4.103)$$

For all the cases,the numerical results indicate that $\bar{g}^h(z) = 5$ for all $z \in (-1,1)$. Hence, for all these cases $\{\zeta_{\alpha}^{\bar{g}}(t), t \ge 0\}$, where $\bar{g}(z) = 5$ for all $z \in (-1,1)$, is a solution to (4.94) (with $g = \bar{g}$), and not merely a weak solution.

Denote by $V_i(\alpha) = V_i(\alpha;\bar{g})$ and $T_i(\alpha) = E_{\alpha}^{\bar{g}} \tau_0(\alpha;\bar{g})$, i=a,b,c,d, the solu= tions respectively for the corresponding systems given by 4.4.1.a - 4.4.1.d. The numerical results showed that $V_b(\alpha) \ge V_i(\alpha)$ and $T_b(\alpha) \ge T_i(\alpha)$ for all $i \in \{a,c,d\}$ and $\alpha \in D_0$. Hence from the point of view of this section, the measurement policy in the case 4.4.1.b is better than the measurement policies of the three other cases.

Figs. 4.1 and 4.2 show the plots of $V(x,0) = V(x,0;\bar{g}^h)$ and $T(x,0) = E_{x,0}^{\bar{g}^h} \tau_0(x,0;\bar{g}^h)$, for the cases 4.4.1.a and 4.4.1.b. The corres= ponding plots of V(x,0) and T(x,0), for the cases 4.4.1.c and 4.4.1.d, turned out to be very similar to Figs. 4.1 and 4.2, and have thus been omitted. Fig. 4.3 shows the plots of V(x,0) and T(x,0) for the case 4.4.1.e.

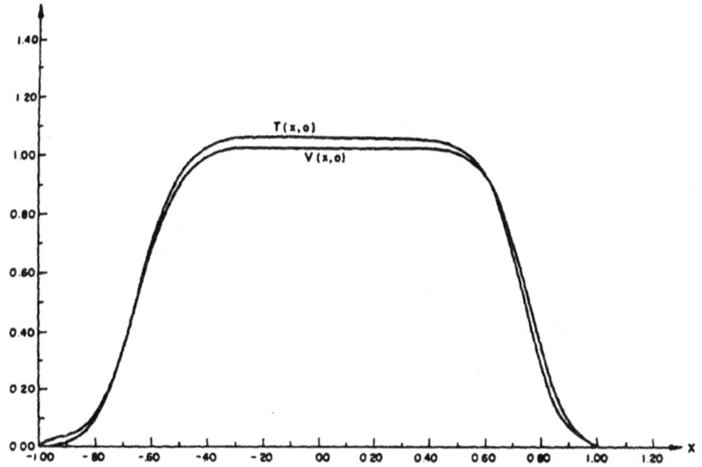

Fig. 4.1: $V(x,0) = V(x,0;\bar{g}^h)$ and $T(x,0) = E_{x,o}^{\bar{g}^h} \tau_o(x,0;\bar{g}^h)$ for the system

$dx = -v_o dt + \sigma dW$, $dy = m_o x dt + \gamma dB$

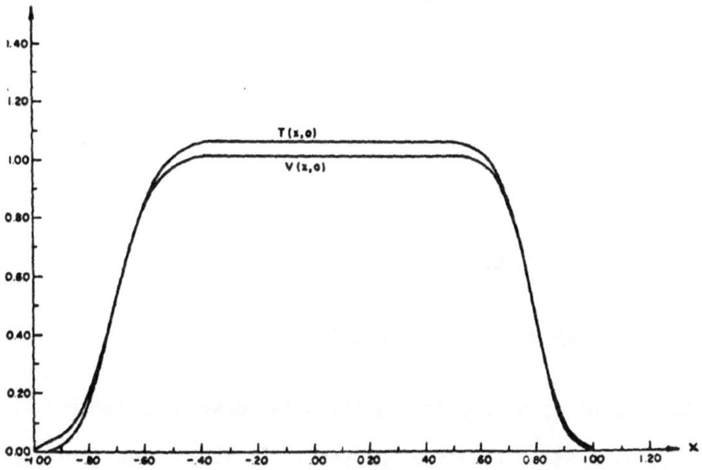

Fig.4.2: $V(x,0) = V(x,0;\bar{g}^h)$ and $T(x,0) = E_{x,o}^{\bar{g}^h} \tau_o(x,0;\bar{g}^h)$ for the system

$dx = -v_o dt + \sigma dW$, $dy = m_o x^3 dt + \gamma dB$

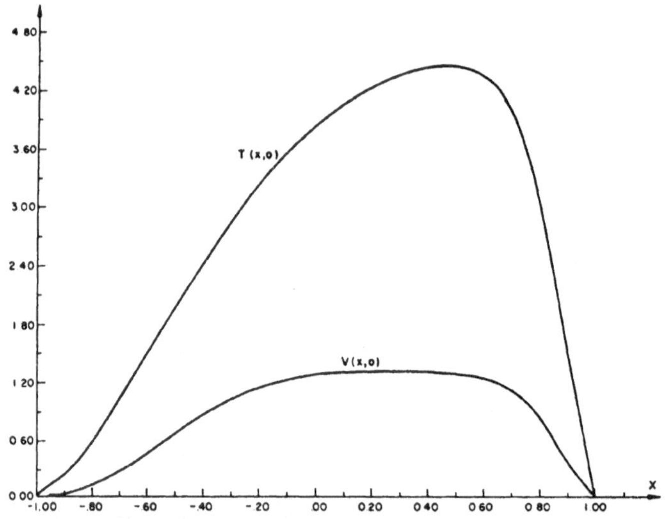

Fig.4.3: $V(x,0) = V(x,0;\bar{g}^h)$ and $T(x,0) = E^{\bar{g}^h}_{x,0}\ \tau_0(x,0;\bar{g}^h)$ for the system $dx = -v_0 dt + \sigma dW$, $dy = m_0 \arctan(x)dt + \gamma dB$.

4.4.2 Example 2

Consider the one-dimensional system given by

$$dx = -v_0 dt + c_0 dN , \quad t > 0 \tag{4.104}$$

with the measurement process given by

$$dy = h(x)dt + \gamma dB , \quad t > 0 \tag{4.105}$$

where $\{N(t), t \geq 0\}$ is a Poisson process with constant parameter λ, and $\{B(t), t \geq 0\}$ is a standard Wiener process. It is assumed here that $\{N(t), t \geq 0\}$ and $\{B(t), t \geq 0\}$ are mutually independent. v_0, c_0, γ and λ are given positive numbers.

The dynamic state estimator is here taken to be of the form

$$dz = [-v_0 + \lambda c_0]dt + g(z)[dy - h(z)dt], \quad t > 0, \tag{4.106}$$

and the sets D_0 and A are given by (4.93) and (4.103) respectively.

Thus equations (4.65) yield

$$\begin{cases} dx = -v_o dt + c_o dN \\ \\ dz = [-v_o + \lambda c_o + g(z)(h(x)-h(z))]dt + \gamma g(z)dB \end{cases} \qquad t > 0 \quad (4.107)$$

and the operators $\mathcal{L}(G)$ (eq. (4.77)) and $\mathcal{L}^*$ (eq. (4.83)) here reduce to

$$\mathcal{L}(g)V(\alpha) = -v_o \partial V(\alpha)/\partial x + [-v_o+\lambda c_o+g(z)(h(x)-h(z))]\partial V(\alpha)/\partial z+(\gamma^2 g_o^2/2)\partial^2 V(\alpha)/\partial z^2 \qquad (4.108)$$

$$+ \lambda[V(x+c_o,z) - V(\alpha)] \quad , V \in \mathcal{D}_o \quad , \alpha \in D_o$$

and

$$\mathcal{L}^*Q(\alpha)=v_o\partial Q(\alpha)/\partial x-(-v_o+\lambda c_o)\partial Q(\alpha)/\partial z+(\gamma^2 g_o^2/2)\partial^2 Q(\alpha)/\partial z^2, \alpha \in D_o \qquad (4.109)$$

for any Q such that $\mathcal{L}^*Q \in L_2(D_o)$.

The functions $V(\alpha) = V(\alpha;\bar{g}^h)$ and $T(\alpha) = E_\alpha^{\bar{g}^h} \tau_o(\alpha;\bar{g}^h)$ were computed for the following systems:

4.4.2.a. $dx = -v_o dt + c_o dN \quad , \quad dy = m_o x dt + \gamma dB$ (4.110)

4.4.2.b. $dx = -v_o dt + c_o dN \quad , \quad dy = m_o x^3 dt + \gamma dB$ (4.111)

4.4.2.c. $dx = -v_o dt + c_o dN \quad , \quad dy = m_o \arctan(x)dt + \gamma dB.$ (4.112)

Computations were carried out for the following set of parameters: $v_o = 0.2$, $c_o = 0.05, 0.1$, $m_o = 10$, $\gamma^2 = 0.05$, $\lambda = 1,2$, $T = 40$, $g_o = 5$ ($G^o=g_o$) and $h = 0.025$. In all cases the numerical results showed that $\bar{g}^h(z)=5$ for all $z \in (-1,1)$, where $(V(\cdot;\bar{g}^h),Q(\cdot;\bar{g}^h),\bar{g}^h)$ is the limit of the sequence $\{(V(\cdot;g^{(n)}), Q(\cdot;g^{(n)}),g^{(n)})\}$ obtained by applying the algorithm for com= puting weak suboptimal gain matrices.

Figs. 4.4, 4.6 and 4.7 show the plots of $V(x,0) = V(x,0;\bar{g}^h)$ and $T(x,0) = E_{x,o}^{\bar{g}^h} \tau_o(x,0;\bar{g}^h)$ for the cases 4.4.2.a, 4.4.2.b and 4.4.2.c res= pectively, where $c_o = 0.05$ and $\lambda = 2$. Fig. 4.5 shows the plots of V(x,0) and T(x,0) for the case 4.4.2.a where $c_o = 0.1$ and $\lambda = 1$.

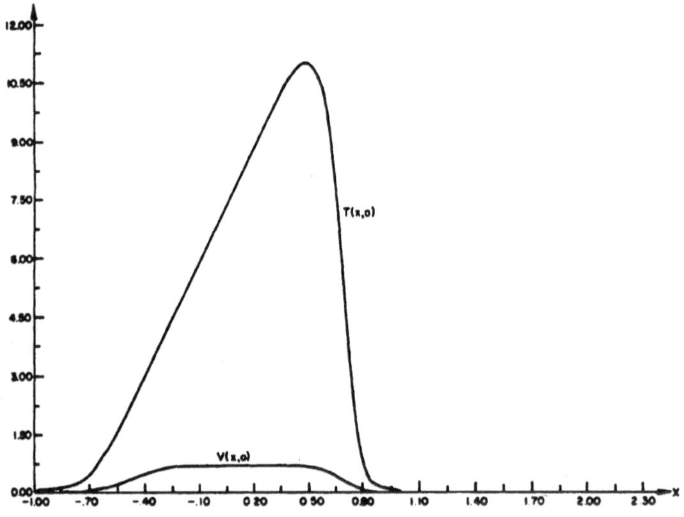

<u>Fig.4.4</u>: $V(x,0) = V(x,0;\bar{g}^h)$ and $T(x,0) = E_{x,o}^{\bar{g}^h} \tau_o(x,0;\bar{g}^h)$ for the system $dx = -v_o dt + c_o dN$, $dy = m_o x dt + \gamma dB$, where $c_o = 0.05$, $\lambda = 2$.

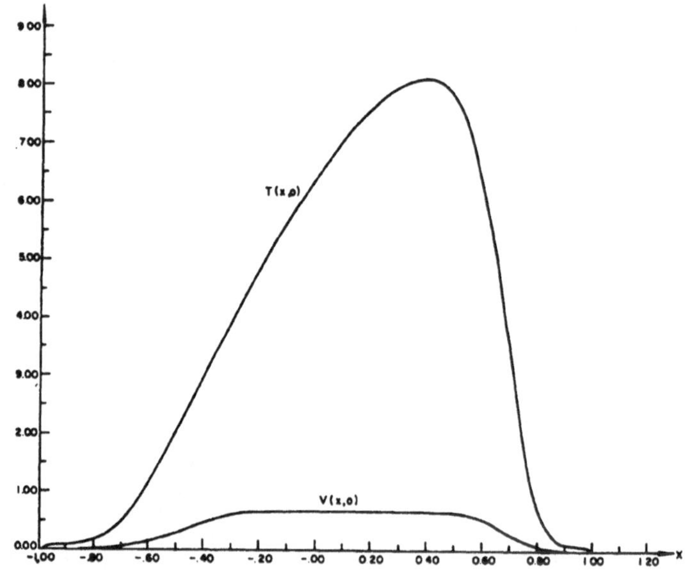

<u>Fig.4.5</u>: $V(x,0) = V(x,0;\bar{g}^h)$ and $T(x,0) = E_{x,o}^{\bar{g}^h} \tau_o(x,0;\bar{g}^h)$ for the system $dx = -v_o dt + c_o dN$, $dy = m_o x dt + \gamma dB$, where $c_o = 0.1$, $\lambda = 1$.

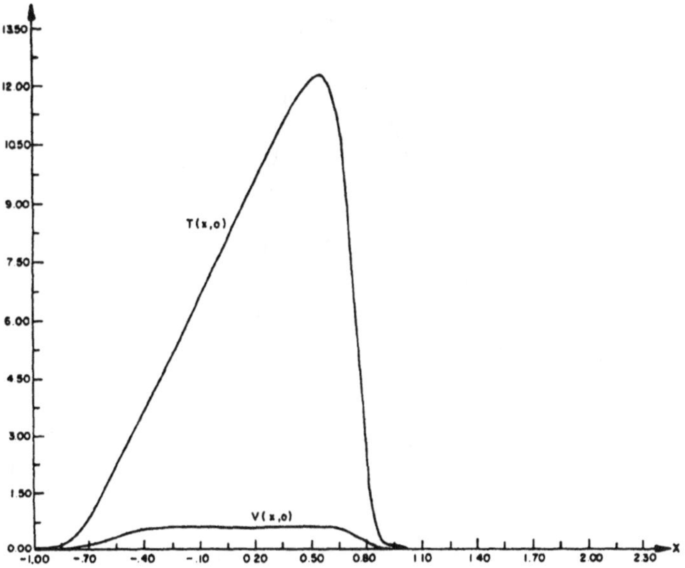

Fig.4.6: $V(x,0) = V(x,0;\bar{g}^h)$ and $T(x,0) = E_{x,o}^{\bar{g}^h}\,\tau_o(x,0;\bar{g}^h)$ for the system
$dx = -v_o dt + c_o dN$, $dy = m_o x^3 dt + \gamma dB$

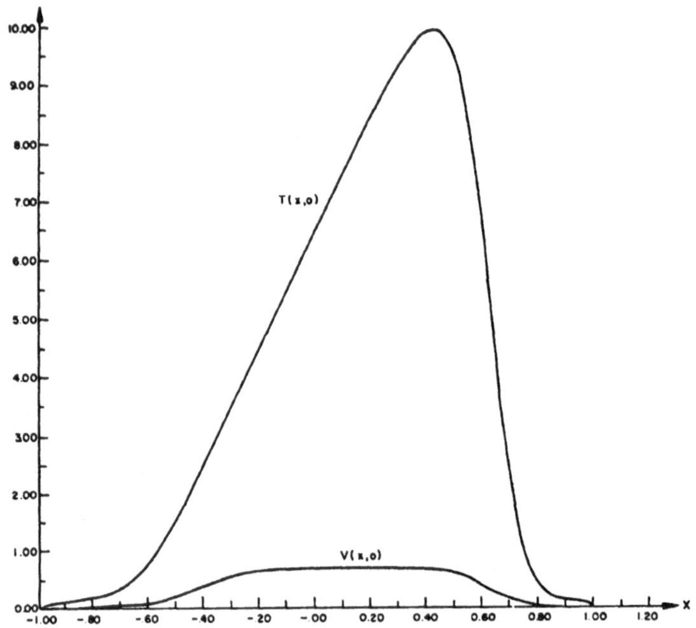

<u>Fig.4.7</u>: $V(x,0) = V(x,0;\bar{g}^h)$ and $T(x,0) = E_{x,o}^{\bar{g}^h}\ \tau_o(x,0;\bar{g}^h)$ for the system

$$dx = -v_o dt + c_o dN\ ,\quad dy = m_o \arctan(x)dt + \gamma dB$$

4.5 AN ALTERNATIVE APPROACH TO NONLINEAR FILTERING: MAXIMIZING THE PRO= BABILITY OF HITTING A TARGET SET

Let a nonlinear stochastic system be given by

$$dx_i = f_i(x)dt + \sigma_i(x)dW_i\ ,\ t > 0,\ x \in \mathbb{R}^m,\ i=1,\dots,m \tag{4.113}$$

$$dy_i = h_i(x)dt + \gamma_i(x)dB_i\ ,\ t > 0,\ i=1,\dots,p\ , \tag{4.114}$$

where $X_t \triangleq \text{col.}(x_1(t),\dots,x_m(t))$, $t \geq 0$ is the state vector, and $Y_t \triangleq \text{col.}(y_1(t),\dots,y_p(t))$, $t \geq 0$ is the output measurement vector, and where $f_i(x)$, $\sigma_i(x)$, $h_j(x)$, $\gamma_j(x)$, $i=1,\dots,m$, $j=1,\dots,p$, $x \in \mathbb{R}^m$ are given functions. $W \triangleq \{W(t) = (W_1(t),\dots,W_m(t)),\ t \geq 0\}$ and $B \triangleq \{B(t) = (B_1(t),\dots,B_p(t)),\ t \geq 0\}$ are $\mathbb{R}^m$-valued and $\mathbb{R}^p$-valued standard Wiener processes respectively.

A dynamic state estimator having the form as given by (4.63) is chosen, and by substituting dy from (4.114) into (4.63), the following set of stochastic differential equations is obtained:

$$\begin{cases} dx_i = f_i(x)dt + \sigma_i(x)dW_i, \quad t > 0, \ i=1,\dots,m \\ dz_i = [f_i(z) + \sum_{j=1}^{p} g_{ij}(z)(h_j(x)-h_j(z))]dt + d\tilde{B}_i, t > 0, i=1,\dots,m \end{cases} \tag{4.115}$$

where

$$d\tilde{B}_i \triangleq \sum_{j=1}^{p} g_{ij}(z)\gamma_j(x)dB_j \quad , \quad i=1,\dots,m. \tag{4.116}$$

It is assumed that f_i, σ_i, h_j and γ_j, $i=1,\dots,m$, $j=1,\dots,p$, satisfy all the conditions stated in Section 4.3. Also, the class $\tilde{U}$, of admissible gain matrices, is defined in the same manner as in Section 4.3. Thus, given $\alpha = (x,z)$ and $G \in \tilde{U}$, equations (4.115) have a unique weak solution $(\zeta_\alpha^G, P_\alpha^G)$, such that $\zeta_\alpha^G(0) = \alpha$, which is a strong Markov process.

Denote by D_o, K and D the following sets in $\mathbb{R}^{2m}$:

$$D_o \triangleq \{\alpha = (x,z) : |x_i| < a_i \text{ and } |z_i| < a_i, \ i=1,\dots,m\}; \tag{4.117}$$

$$K \triangleq \{\alpha = (x,z) : |x-z| \le \varepsilon , \ |x_i| \le a_i-\delta \text{ and } |z_i| \le a_i-\delta, \ i=1,\dots,m\}; \tag{4.118}$$

$$D \triangleq D_o - K; \tag{4.119}$$

where $a_i > 0, i=1,\dots,m$, $\varepsilon > 0$ and $0 < \delta << \min_i a_i$ are given numbers. Define for $G \in \tilde{U}$,

$$\tau(\alpha;G) \triangleq \begin{cases} \inf\{t : \zeta_\alpha^G(t) \notin D \text{ when } \zeta_\alpha^G(0) = \alpha \in D\} \\ 0 \qquad \text{if } \zeta_\alpha^G(0) = \alpha \notin D \\ \infty \qquad \text{if } \zeta_\alpha^G(t) \in D \text{ for all } t \ge 0. \end{cases} \tag{4.120}$$

Hence τ is the first exit time of ζ_α^G from D. Also,let

$$U \triangleq \{G \in \tilde{U} : \sup_{\alpha \in D} E_\alpha^G \tau(\alpha;G) < \infty\} , \tag{4.121}$$

and define

$$V(\alpha;G) \triangleq P_\alpha^G(\{\zeta_\alpha^G(\tau(\alpha;G)) \in K\}), \; G \in U \tag{4.122}$$

i.e. $V(\alpha;G)$ is the probability that ζ_α^G enters the set K before hitting the set ∂D_o (∂D_o denotes the boundary of D_o).

Denote

$$\ell(G) \triangleq \int_{D_o} (1 - V(\alpha;G))^2 d\alpha. \tag{4.123}$$

In this section the following estimation problem is treated: Find a gain matrix $G^* \in U$ such that

$$\ell(G^*) \leq \ell(G) \text{ for any } G \in U. \tag{4.124}$$

Let $\mathcal{D}_o$ denote the class of all functions $V = V(\alpha)$ such that : V is continuous on the closure $\bar{D}_o$ of D_o and twice continuously differentiable on D; for any $G \in U$, $\mathcal{L}(G)V \in L_2(D_o)$ where

$$\mathcal{L}(G)V(\alpha) \triangleq \sum_{i=1}^{m} \{f_i(x) \frac{\partial V(\alpha)}{\partial x_i} + [f_i(z) + \sum_{j=1}^{p} g_{ij}(z)(h_j(x)-h_j(z))] \frac{\partial V(\alpha)}{\partial z_i}\} + \frac{1}{2} \sum_{i,j=1}^{2m} (\sigma(x,G^o)\sigma'(x,G^o))_{ij} \frac{\partial^2 V(\alpha)}{\partial \alpha_i \partial \alpha_j} \tag{4.125}$$

and $x = (\alpha_1,\ldots,\alpha_m)$, $z = (\alpha_{m+1},\ldots,\alpha_{2m})$.

Again, as in Section 4.3, the operator $\mathcal{L}(G)$ is the infinitesimal gene= rator of the family of Markov processes $\{(\zeta_\alpha^G,P_\alpha^G), \alpha \in \bar{D}_o\}$.

Given $G \in U$. Let $V \in \mathcal{D}_o$ satisfy

$$\mathcal{L}(G)V(\alpha) = 0, \; \alpha \in D \; ; \;\; V(\alpha) = 1, \; \alpha \in K \; ; \;\; V(\alpha) = 0 \; , \; \alpha \notin D_o \tag{4.126}$$

then, it can be shown that

$$V(\alpha) = V(\alpha;G) = P_\alpha^G(\{\zeta_\alpha^G(\tau(\alpha;G)) \in K\}), \; \alpha \in \bar{D}_o \tag{4.127}$$

In the same manner as in Section 2.6, we define the set U_o

$$U_o \triangleq \{G \in U : V(\alpha;G) \in \mathcal{D}_o \text{ and satisfies eqns (4.126)}\}, \qquad (4.128)$$

and the operator $\mathcal{L}^*$

$$\mathcal{L}^*Q(\alpha) \triangleq - \sum_{i=1}^{m} \{\partial[f_i(x)Q(\alpha)]/\partial x_i + \partial[f_i(z)Q(\alpha)]/\partial z_i\}$$
$$+ (\tfrac{1}{2}) \sum_{i,j=1}^{2m} \partial^2[(\sigma(x,G^o)\sigma'(x,G^o))_{ij}Q(\alpha)]/\partial\alpha_i\partial\alpha_j \qquad (4.129)$$

for all Q such that $\mathcal{L}^*Q \in L_2(D_o)$.

Applying the same procedure as in the proof of Theorem 2.3, the following theorem is obtained.

Theorem 4.3

Suppose $V_o \in \mathcal{D}_o$, $G^* \in U_o$ and Q_o satisfy

$$\mathcal{L}(G^*)V_o(\alpha) = 0,\ \alpha \in D;\ V_o(\alpha) = 1,\ \alpha \in K;\ V_o(\alpha) = 0,\ \alpha \notin D_o\ , \qquad (4.130)$$

$$\mathcal{L}^*Q_o(\alpha) = 1 - V_o(\alpha),\ \text{a.e. in } D_o\ ;\ Q_o(\alpha) = 0,\ \alpha \notin D_o, \qquad (4.131)$$

where $G^* \in U_o$ is determined by

$$G^* = \arg\sup_{G \in U_o} \{- \sum_{i=1}^{m} \sum_{j=1}^{p} \int_{D_{oz}} g_{ij}(z) \int_{D_{ox}} Q_o(\alpha)[h_j(x)-h_j(z)](\partial V(\alpha;G)/\partial z_i)dxdz\} \qquad (4.132)$$

and

$$D_{ox} \triangleq \{x : |x_i| < a_i,\ i=1,\dots,m\},\quad D_{oz} \triangleq \{z : |z_i| < a_i,\ i=1,\dots,m\}; \qquad (4.133)$$

then

$$\ell(G^*) \le \ell(G) \quad \text{for any } G \in U_o. \qquad (4.134)$$

A procedure for computing weak suboptimal gain matrices, similar to that given in Section 4.3, can be applied to equations (4.130)-(4.132).

4.6 NUMERICAL EXAMPLES

The following examples are taken from Yavin [80].

4.6.1 Example 1

Consider the one-dimensional system

$$dx = f(x)dt + \sigma dW \ , \quad t > 0 \tag{4.135}$$

with the measurement process $\{Y_t, t \geq 0\}$ determined by

$$dy = h(x)dt + \gamma dB \quad , \quad t > 0 \quad , \quad y \in \mathbb{R}, \tag{4.136}$$

where $f(x)$ and $h(x)$ are given functions; σ and γ are given positive num= bers; and $\{W(t), t \geq 0\}$ and $\{B(t), t \geq 0\}$ are two independent one-dimen= sional standard Wiener processes. In this case the dynamic state estima= tor is given by

$$dz = f(z)dt + g(z)(dy - h(z)dt), \ t > 0, \ z \in \mathbb{R}, \tag{4.137}$$

and the set D_0 is here chosen to be

$$D_0 \triangleq \{\alpha = (x,z) : |x| < 1 \text{ and } |z| < 1\}. \tag{4.138}$$

The operators $\mathcal{L}(G)$, (4.125) and $\mathcal{L}^*$, (4.129), here reduce to

$$\mathcal{L}(g)V(\alpha) = f(x)\partial V(\alpha)/\partial x + [f(z) + g(z)(h(x) - h(z))]\partial V(\alpha)/\partial z$$

$$+ (\tfrac{1}{2})[\sigma^2 \ \partial^2 V(\alpha)/\partial x^2 + g_0^2 \ \gamma^2 \ \partial^2 V(\alpha)/\partial z^2], \ \alpha \in D, \ V \in \mathcal{D}_0 \tag{4.139}$$

$$\mathcal{L}^*Q(\alpha) = -\partial[f(x)Q(\alpha)]/\partial x - \partial[f(z)Q(\alpha)]/\partial z + (\tfrac{1}{2})[\sigma^2\partial^2 Q(\alpha)/\partial x^2 + g_0^2\gamma^2\partial^2 Q(\alpha)/\partial z^2] \tag{4.140}$$

for any Q such that $\mathcal{L}^*Q \in L_2(D_0)$,

where $G^0 = g_0$ and g_0 is a given positive number.

The procedure for computing weak suboptimal gain matrices, where $g^{(n+1)}$ is determined by

$$g^{(n+1)}(z) = -g_0 \text{sign}(\int_{-1}^{1} Q(\alpha;g^{(n)})(h(x)-h(z))(\partial V(\alpha;g^{(n)})/\partial z)dx), \; z \in (-1,1) \tag{4.141}$$

$n=0,1,\ldots$;

has been applied for the following cases:

(a) $dx = -v_0 dt + \sigma dW, \; dy = h_0 \text{ arc } \tan(x)dt + \gamma dB, \; t > 0;$ (4.142)

(b) $dx = v_0 x dt + \sigma dW, \; dy = h_0 x^3 dt + \gamma dB \;, \; t > 0;$ (4.143)

(c) $dx = a_0 x^3 dt + \sigma dW, \; dy = h_0 x dt + \gamma dB, \; t > 0;$ (4.144)

(d) $dx = a_0 x^5 dt + \sigma dW, \; dy = h_0 x dt + \gamma dB, \; t > 0;$ (4.145)

(e) $dx = a_0 x^7 dt + \sigma dW, \; dy = h_0 x dt + \gamma dB, \; t > 0;$ (4.146)

where $v_0 = 0.1, 0.2, 0.4, 0.8$; $h_0 = 10$; $g_0 = 5$; $\sigma^2 = 0.04$; $\gamma^2 = 0.01$; $\varepsilon = 0.01$; $a_0 = 0.2$ and $h = 0.025$.

In all the cases solved, the algorithm for computing weak suboptimal gain matrices led to fast convergence of $\{g^{(n)}\}$ to $\bar{g}^h$. For all the cases, the numerical results indicate that $\bar{g}^h(z) = g_0$ for all $z \in (-1,1)$.

Fig. 4.8 shows the plot of $V(x,0;\bar{g}^h)$ for case (b) where $v_0 = 0.1$, and Fig. 4.9 shows the plot of $V(x,0;\bar{g}^h)$ for case (c). The plots of $V(x,0;\bar{g}^h)$ for the rest of the cases turned out to be very similar either to Fig. 4.8 or Fig. 4.9 and have thus been omitted.

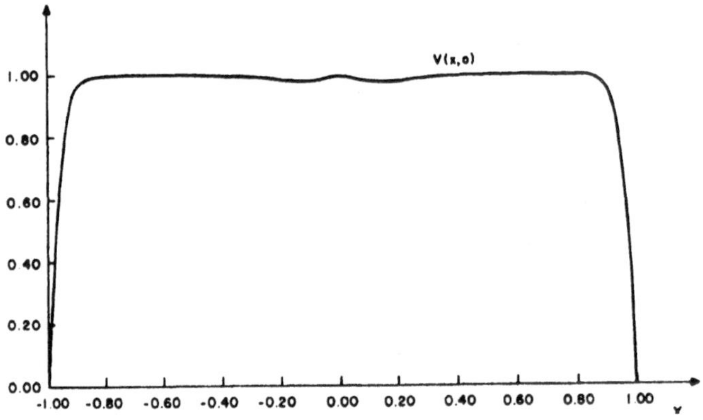

Fig. 4.8: The plot of $V(x,0;\bar{g}^h)$ as a function of x, for $f(x) = 0.1x$, $h(x) = 10x^3$, $\sigma^2 = 0.04$, $\gamma^2 = 0.01$, $\varepsilon = 0.01$, $g_o = 5$ and $h = 0.025$.

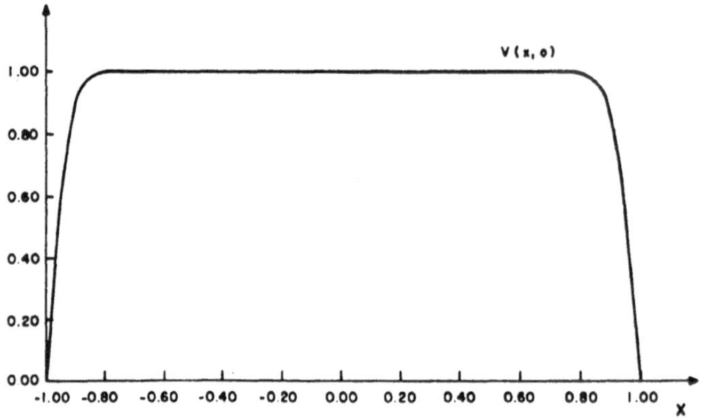

Fig.4.9: The plot of $V(x,0;\bar{g}^h)$ as a function of x, for $f(x) = 0.2x^3$, $h(x) = 10x$, $\sigma^2 = 0.04$, $\gamma^2 = 0.01$, $\varepsilon = 0.01$, $g_o = 5$ and $h = 0.025$

4.6.2 Example 2

Consider the one-dimensional system given by (4.135) with the two-dimensional measurement process $\{Y_t = (y_1(t), y_2(t)),\ t \geq 0\}$ determined by

$$\begin{cases} dy_1 = \sqrt{h_o^2 + x^2}\,dt + \gamma_1 dB_1 \\ \qquad\qquad\qquad\qquad\qquad\qquad t > 0, \\ dy_2 = \arctan(x/h_o)dt + \gamma_2 dB_2 \end{cases} \tag{4.147}$$

where $\{W(t),\ t \geq 0\}$, $\{B_1(t),\ t \geq 0\}$ and $\{B_2(t),\ t \geq 0\}$ are mutually independent standard Wiener processes, and σ, h_o, γ_1 and γ_2 are given positive numbers. In this case the dynamic state estimator is given by

$$\begin{aligned} dz = {} & f(z)dt + g_1(z)(dy_1 - \sqrt{h_o^2 + z^2}\,dt) \\ & + g_2(z)(dy_2 - \arctan(z/h_o)dt),\ t > 0. \end{aligned} \tag{4.148}$$

The set D_o is here given by (4.138).

Thus, steps 2,4 and 5 of the algorithm for computing weak suboptimal gain matrices, here reduce to

$$\begin{cases} f(x)\dfrac{\partial V(\alpha)}{\partial x} + [f(z) + g_1^{(n)}(z)(\sqrt{h_o^2 + x^2} - \sqrt{h_o^2 + z^2}) \\ \qquad + g_2^{(n)}(z)(\arctan(x/h_o) - \arctan(z/h_o))]\dfrac{\partial V(\alpha)}{\partial z} \\ \qquad + \tfrac{1}{2}\{\sigma^2 \dfrac{\partial^2 V(\alpha)}{\partial x^2} + [\gamma_1^2(g_1^o)^2 + \gamma_2^2(g_2^o)^2]\dfrac{\partial^2 V(\alpha)}{\partial z^2}\} = 0,\ \alpha \in D \\ \qquad V(\alpha) = 1,\ \alpha \in K;\quad V(\alpha) = 0,\ \alpha \notin D_o. \end{cases} \tag{4.149}$$

$$\begin{cases} -\dfrac{\partial}{\partial x}[f(x)Q(\alpha)] - \dfrac{\partial}{\partial z}[f(z)Q(\alpha)] \\ \quad + \tfrac{1}{2}\{\sigma^2\dfrac{\partial^2 Q(\alpha)}{\partial x^2} + [\gamma_1^2(g_1^o)^2 + \gamma_2^2(g_2^o)^2]\dfrac{\partial^2 Q(\alpha)}{\partial z^2}\} = 1 - V(\alpha; G^{(n)}),\ \alpha \in D_o \\ Q(\alpha) = 0\ ,\ \alpha \notin D_o\ . \end{cases} \tag{4.150}$$

$$\begin{cases} g_1^{(n+1)}(z) = - g_1^o \, \mathrm{sign}\{\int_{-1}^{1} Q(\alpha;G^{(n)})(\sqrt{h_o^2+x^2} - \sqrt{h_o^2+z^2}) \frac{\partial V(\alpha;G^{(n)})}{\partial z} dx\} \\ g_2^{(n+1)}(z) = - g_2^o \, \mathrm{sign}\{\int_{-1}^{1} Q(\alpha;G^{(n)})(\arctan(x/h_o) - \arctan(z/h_o)) \frac{\partial V(\alpha;G^{(n)})}{\partial z} dx\} \end{cases} \tag{4.151}$$

$n=0,1,2,\ldots$

respectively, where $G^{(n)} = (g_1^{(n)}, g_2^{(n)})$ and g_i^o, $i=1,2$, are given positive numbers.

The procedure for computing weak suboptimal gain matrices has been applied for the following cases:

(a) $$dx = a_o x dt + \sigma dW \, , \; t > 0; \tag{4.152}$$

(b) $$dx = a_o x^3 dt + \sigma dW \, , \; t > 0; \tag{4.153}$$

where $a_o = 0.2$; $h_o = 1$; $g_1^o = 10,5$; $g_2^o = 5$; $\sigma^2 = 0.04$; $\gamma_1^2(g_1^o)^2 + \gamma_2^2(g_2^o)^2 = 1.125$, 5.125; $h = 0.025, 0.0125$; $\varepsilon = 0.01, 0.025, 0.0125$.

In all the cases solved, the algorithm for computing weak suboptimal gain matrices led to fast convergence of $\{G^{(n)}\}$ to $\bar{G}^h$. The numerical re= sults indicate the following possible forms for $\bar{g}_1^h$:

$$\bar{g}_1^h(z) = \begin{cases} g_1^o \, \mathrm{sign}(z_1) \text{ if : } g_1^o = g_2^o = 5, \; \gamma_1^2(g_1^o)^2 + \gamma_2^2(g_2^o)^2 = 5.125 \\ g_1^o \, \mathrm{sign}(z_1 - 0.25) \text{ if : } g_1^o = g_2^o = 5, \; \gamma_1^2(g_1^o)^2 + \gamma_2^2(g_2^o)^2 = 1.125 \\ g_1^o \, \mathrm{sign}(z_1 + 0.025) \text{ if: } g_1^o = 10, \; g_2^o = 5, \; \gamma_1^2(g_1^o)^2 + \gamma_2^2(g_2^o)^2 = 5.125 \end{cases} \tag{4.154}$$

$z \in (-1,1)$

while

$$\bar{g}_2^h(z) = g_2^o \quad z \in (-1,1) \tag{4.155}$$

Figs. 4.10-4.12 show plots of $V(x,0;\bar{G}^h)$ for three typical cases.

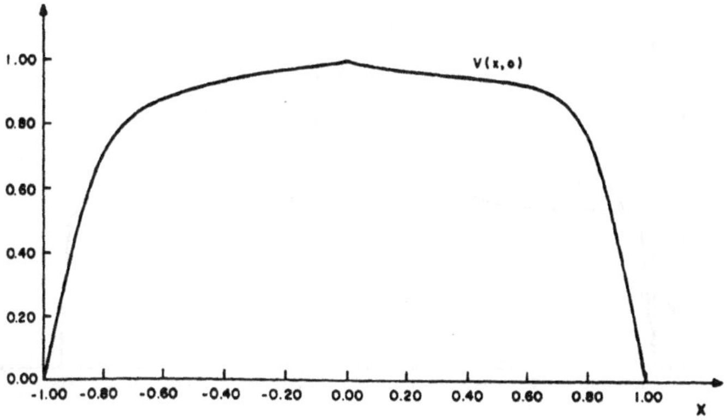

Fig.4.10: The plot of $V(x,0;\bar{G}^h)$ as a function of x, for $f(x) = 0.2x$, $h_1(x) = \sqrt{1+x^2}$, $h_2(x) = \arctan(x)$, $\sigma^2 = 0.04$, $\gamma_1^2(g_1^0)^2 + \gamma_2^2(g_2^0)^2 = 1.125$, $\varepsilon = 0.01$, $g_1^0 = g_2^0 = 5$ and $h = 0.025$.

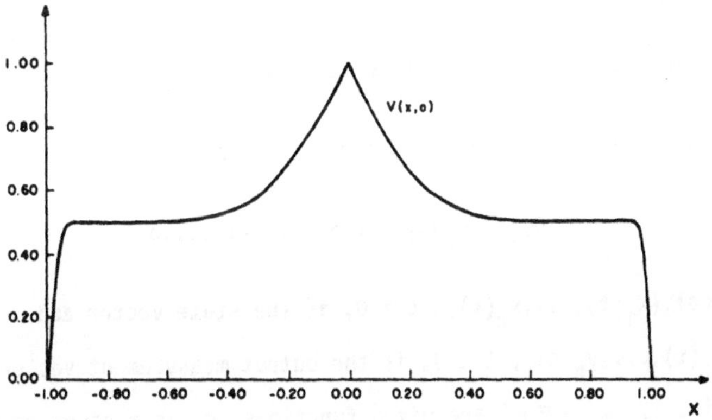

Fig.4.11: The plot of $V(x,0;\bar{G}^h)$ as a function of x, for $f(x) = 0.2x^3$, $h_1(x) = \sqrt{1+x^2}$, $h_2(x) = \arctan(x)$, $\sigma^2 = 0.04$, $\gamma_1^2(g_1^0)^2 + \gamma_2^2(g_2^0)^2 = 5.125$, $\varepsilon = 0.01$, $g_1^0 = g_2^0 = 5$ and $h = 0.0125$.

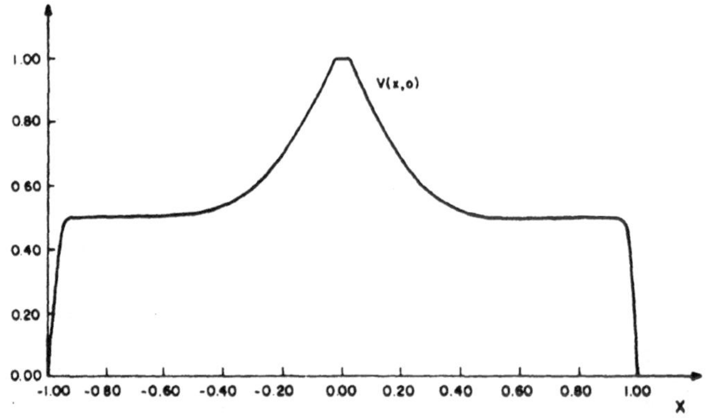

Fig.4.12: The plot of $V(x,0;\bar{G}^h)$ as a function of x, for $f(x) = 0.2x^3$, $h_1(x) = \sqrt{1+x^2}$, $h_2(x) = \arctan(x)$, $\sigma^2 = 0.04$, $\gamma_1^2(g_1^0)^2 + \gamma_2^2(g_2^0)^2 = 5.125$, $\varepsilon = 0.025$, $g_1^0 = g_2^0 = 5$ and $h = 0.0125$.

4.7 AN ALTERNATIVE APPROACH TO NONLINEAR FILTERING: JUMP PROCESS OBSERVATIONS

4.7.1 Introduction

Let a nonlinear stochastic system be given by

$$dx_i = f_i(x)dt + \sigma_i(x)dW_i \ , \ t > 0 \ , \ x \in \mathbb{R}^m, \ i=1,\dots,m \tag{4.156}$$

$$dy_i = c_0 dN_i \ , \ t > 0 \ , \ i=1,\dots,p \tag{4.157}$$

where $X_t \triangleq \text{col.}(x_1(t),\dots,x_m(t))$, $t \geq 0$, is the state vector and $Y_t = \text{col.}(y_1(t),\dots,y_p(t))$, $t \geq 0$, is the output measurement vector. $f_i(x)$ and $\sigma_i(x)$, $i=1,\dots,m$, $x \in \mathbb{R}^m$ are given functions, c_0 is a given positive number, and $\{W(t) \triangleq (W_1(t),\dots,W_m(t)),\ t \geq 0\}$ is an $\mathbb{R}^m$-valued standard Wiener process. Two cases are dealt with here.

Case 1: N_i, i=1,...,p are square integrable counting processes for all $t \geq 0$. Let $F_t = \sigma\{X_s, N_s, s \leq t\}$ (F_t is the σ-algebra generated by $\{X_s, s \leq t\}$ and $\{N_s, s \leq t\}$) and denote by F the σ-algebra generated by the union of $(F_t, t \geq 0)$. On a probability space (Ω, F, P) it is assumed here that W is an (F_t, P) $\mathbb{R}^m$-valued Wiener process, and that there is a given function $\lambda : \mathbb{R}^m \to \mathbb{R}^p$ such that λ_i, i=1,...,p are continuously differentiable on $\mathbb{R}^m$; $\lambda_i(x) \geq \lambda_0 > 0$, i=1,...,p, $x \in \mathbb{R}^m$; and that

$$M(t) \triangleq N(t) - \int_0^t \lambda(X_s)ds \quad , \quad t \geq 0 \tag{4.158}$$

is a local (F_t, P) square integrable $\mathbb{R}^p$-valued martingale. For an exposition of martingales and related processes see Meyer [108], or Liptser and Shiryayev [56], or Gihman and Skorohod [2], or Boel et al. [107].

Case 2: In this case $c_0 = 1$, and $\{N_i\}$ are given by

$$N_i(t) = \int_0^t \int_{\mathbb{R}^m} c_i(X_s, u)\nu(ds, du) \quad , \quad t \geq 0 \ , \ i=1,\ldots,p \tag{4.159}$$

where $\{\nu(t,A); t \geq 0\}$, $A \in B(\mathbb{R}^m)$, is a Poisson process with

$$E\nu(t,A) = t\pi(A) \quad , \quad t \geq 0, \ A \in B(\mathbb{R}^m). \tag{4.160}$$

$B(\mathbb{R}^m)$ denotes the m-dimensional Borel σ-algebra and $c_i(x,u)$, i=1,...,p, $x,u \in \mathbb{R}^m$ are given functions. It is assumed here that π is a given measure on $B(\mathbb{R}^m)$, $\pi(\mathbb{R}^m) < \infty$, and that the processes $\{W(t), t \geq 0\}$, $\{\nu(t,A), t \geq 0\}$ are mutually independent for any $A \in B(\mathbb{R}^m)$. For more details on $\{\nu(t,A), t \geq 0, A \in B(\mathbb{R}^m)\}$ see Gihman and Skorohod [1].

For the system given by eqns (4.156)-(4.157), a dynamic state estimator is chosen having the following form

$$\left.\begin{aligned} dz_i &= f_i(z)dt + \sum_{j=1}^{p} g_{ij}(y,z)(dy_j - c_0\lambda_j(z)dt) \\ t &> 0, \ i=1,\ldots,m \quad , \quad z \in \mathbb{R}^m \quad , \quad y \in \mathbb{R}^p \end{aligned}\right\} \text{ in case 1} \tag{4.161}$$

or

$$\left.\begin{aligned} &dz_i = f_i(z)dt + \sum_{j=1}^{p} g_{ij}(y,z)(dy_j - c_j(z)dt),\ t > 0,\ i=1,\dots,m \\ &z \in \mathbb{R}^m\ ,\ y \in \mathbb{R}^p \end{aligned}\right\}\text{in case 2} \qquad (4.162)$$

where

$$c_i(x) \triangleq \int_{\mathbb{R}^m} c_i(x,u)\pi(du)\ ,\ i=1,\dots,p,\ x \in \mathbb{R}^m\ , \qquad (4.163)$$

and where in both cases the functions $\{g_{ij}\}$ are yet to be determined. The matrix

$$G(y,z) \triangleq \begin{pmatrix} g_{11}(y,z) & \dots & g_{1p}(y,z) \\ \vdots & & \\ g_{m1}(y,z) & \dots & g_{mp}(y,z) \end{pmatrix} \qquad (4.164)$$

will here be called the *gain matrix of the observer* (4.161) (or (4.162)). The dynamic system given by eqn (4.161) (or eqn (4.162)) uses y from eqn (4.157) as input, and its output serves as an estimate of the state of the system.

By substituting dy from eqn (4.157) into eqn (4.161) (or eqn (4.162)), the following sets of stochastic differential equations are obtained:

$$\left.\begin{aligned} &dx_i = f_i(x)dt + \sigma_i(x)dW_i\ ,\ t > 0\ ,\ i=1,\dots,m \\ &dy_i = c_o\ \lambda_i(x)dt + c_o dM_i\ ,\ t > 0\ ,\ i=1,\dots,p \\ &dz_i = [f_i(z) + c_o \sum_{j=1}^{p} g_{ij}(y,z)(\lambda_j(x) - \lambda_j(z))]dt + c_o d\eta_i, \\ &t > 0,\ i=1,\dots,m \end{aligned}\right\}\text{in case 1} \qquad (4.165)$$

where

$$d\eta_i \triangleq \sum_{j=1}^{p} g_{ij}(y,z)dM_j\ ,\ t > 0\ ,\ i=1,\dots,m \qquad (4.166)$$

or

$$\left.\begin{aligned} dx_i &= f_i(x)dt + \sigma_i(x)dW_i \;,\quad t > 0 \;,\; i=1,\dots,m \\ dy_i &= c_i(x)dt + \int_{\mathbb{R}^m} c_i(x,u)q(dt,du),\; t > 0, i=1,\dots,p \\ dz_i &= [f_i(z) + \sum_{j=1}^{p} g_{ij}(y,z)(c_j(x)-c_j(z))]dt + dB_i, \end{aligned}\right\} \text{in case 2} \quad (4.167)$$

$t > 0,\; i=1,\dots,m$

where

$$q(t,A) = \nu(t,A) - t\pi(A),\; t \geq 0,\; A \in \mathcal{B}(\mathbb{R}^m) \tag{4.168}$$

and

$$dB_i \triangleq \sum_{j=1}^{p} g_{ij}(y,z) \int_{\mathbb{R}^m} c_j(x,u)q(dt,du),\; t > 0,\; i=1,\dots,m. \tag{4.169}$$

Here and in the sequel, eqns (4.165) and (4.167) have to be interpre= ted as shorthand notations for the following sets of stochastic integral equations respectively:

$$\left.\begin{aligned} x_i(t) &= x_i + \int_0^t f_i(X_{s-})ds + \int_0^t \sigma_i(X_{s-})dW_i(s),\; t \geq 0,\; i=1,\dots,m \\ y_i(t) &= y_i + c_o \int_0^t \lambda_i(X_{s-})ds + c_o M_i(t),\; t \geq 0,\; i=1,\dots,p \\ z_i(t) &= z_i + \int_0^t [f_i(Z_{s-}) + c_o \sum_{j=1}^{p} g_{ij}(Y_{s-},Z_{s-})(\lambda_j(X_{s-})-\lambda_j(Z_{s-}))]ds \\ &+ c_o \int_0^t \sum_{j=1}^{p} g_{ij}(Y_{s-},Z_{s-})dM_j(s) \;,\; t \geq 0,\; i=1,\dots,m \end{aligned}\right\} (4.170)$$

in case 1, and

$$\left.\begin{aligned} x_i(t) &= x_i + \int_0^t f_i(X_{s-})ds + \int_0^t \sigma_i(X_{s-})dW_i(s),\; t \geq 0,\; i=1,\dots,m \\ y_i(t) &= y_i + \int_0^t c_i(X_{s-})ds + \int_0^t \int_{\mathbb{R}^m} c_i(X_{s-},u)q(ds,du), t \geq 0, i=1,\dots,p \\ z_i(t) &= z_i + \int_0^t [f_i(Z_{s-}) + \sum_{j=1}^{p} g_{ij}(Y_{s-},Z_{s-})(c_j(X_{s-})-c_j(Z_{s-}))]ds \\ &+ \int_0^t \int_{\mathbb{R}^m} \sum_{j=1}^{p} g_{ij}(Y_{s-},Z_{s-})c_j(X_{s-},u)q(ds,du) \end{aligned}\right\} (4.171)$$

$t \geq 0,\; i=1,\dots,m$

in case 2, where in (4.170) and (4.171), $Z_t = (z_1(t),\dots,z_m(t))$, $t \geq 0$.

It is assumed here that f_i, $i=1,\dots,m$ are bounded and continuously differentiable on $\mathbb{R}^m$; that σ_i, $i=1,\dots,m$ are bounded and twice continuously differentiable on $\mathbb{R}^m$; and that c_i, $i=1,\dots,p$ satisfy the following conditions:

$$\int_{\mathbb{R}^m} c_i^2(x,u)\pi(du) \leq \ell_o(1 + |x|^2), \ i=1,\dots,p, \ x \in \mathbb{R}^m \tag{4.172}$$

and

$$\int_{\mathbb{R}^m} (c_i(x,u) - c_i(x',u))^2\pi(du) \leq \ell_o|x-x'|^2, \ i=1,\dots,p; x,x' \in \mathbb{R}^m \tag{4.173}$$

for $\ell_o < \infty$ and that $c_i(x)$, $i=1,\dots,p$, are continuously differentiable on $\mathbb{R}^m$.

Let $\tilde{U}$ denote the class of all gain matrices $G = G(y,z)$, of bang-bang type such that : (i) g_{ij} is measurable and $|g_{ij}(y,z)| = g_{ij}^o$ for $i=1,\dots,m$ $j=1,\dots,p$ and all $(y,z) \in \mathbb{R}^p \times \mathbb{R}^m$, where $\{g_{ij}^o\}$ are given non-negative numbers. (ii) $\{(\eta_1(t),\dots,\eta_m(t)), t \geq 0\}$, (4.166), is a local (F_t,P) square integrable martingale (in case 1); or $\{(B_1(t),\dots,B_m(t)), t \geq 0\}$, (4.169), is a local (F_t,P) square integrable martingale (in case 2). In case 2, $F_t = \sigma(X_s,\nu(s,A), s \leq t, A \in B(\mathbb{R}^m))$.

Given $G \in \tilde{U}$, denote by $\zeta_\alpha^G(t) = (X_t,Y_t,Z_t^G)$ the solution to eqn (4.170) and by $\chi_\alpha^G(t) = (X_t,Y_t,Z_t^G)$ the solution to eqn (4.171), where $\zeta_\alpha^G(0) = \chi_\alpha^G(0) = \alpha = (x,y,z)$.

Denote by D the following open set in $\mathbb{R}^{2m+p}$:

$$D \triangleq \{\alpha = (x,y,z): |x_i| < 1, \ |y_j| < 1, \ |z_i| < 1, i=1,\dots,m, j=1,\dots,p\} \tag{4.174}$$

and let D^c denote the complement of D. Define, for $G \in \tilde{U}$

$$\tau_1(\alpha;G) \triangleq \begin{cases} \inf\{t : \zeta_\alpha^G(t) \in D^c \text{ when } \zeta_\alpha^G(0) = \alpha \in D\} & \\ 0 & \text{if } \zeta_\alpha^G(0) = \alpha \in D^c \\ \infty & \text{if } \zeta_\alpha^G(t) \in D \text{ for all } t \geq 0 \end{cases} \tag{4.175}$$

$$\tau_2(\alpha;G) \triangleq \begin{cases} \inf\{t : \chi_\alpha^G(t) \in D^c \text{ when } \chi_\alpha^G(0) = \alpha \in D\} & \\ 0 & \text{if } \chi_\alpha^G(0) = \alpha \in D^c \\ \infty & \text{if } \chi_\alpha^G(t) \in D \text{ for all } t \geq 0. \end{cases} \tag{4.176}$$

Let

$$U_i \triangleq \{G \in \tilde{U} : \sup_{\alpha \in D} E_\alpha \, \tau_i(\alpha;G) < \infty\} \ , \ i=1,2 \ , \tag{4.177}$$

and for a given positive number ε, held fixed, set

$$V_i(\alpha;G) \triangleq E_\alpha \, \Lambda\{t : 0 \leq t < \tau_i(\alpha;G), \ |X_t - Z_t^G| \leq \varepsilon\}, \ G \in U_i, i=1,2 \tag{4.178}$$

where $E_\alpha = E[\cdot|\zeta_\alpha^G(0) = \alpha]$ for i=1 and $E_\alpha = E[\cdot|\chi_\alpha^G(0) = \alpha]$ for i=2 (and Λ is the Lebesgue measure on the real line). In the definition of V_1, X_t and Z_t^G are components of $\zeta_\alpha^G(t)$, and in the definition of V_2, X_t and Z_t^G are components of $\chi_\alpha^G(t)$.

Following the discussion of Section 4.3, the functionals $\ell_i(G)$, i=1,2, are defined as follows:

$$\ell_i(G) \triangleq \int_D (T - V_i(\alpha;G))^2 d\alpha, \ G \in U_i, \ \ i=1,2, \tag{4.179}$$

where $0 < T < \infty$ is a given number satisfying

$$\sup_{G \in U_i} E_\alpha \, \tau_i(\alpha;G) < T \text{ for all } \alpha \in D, \ i=1,2. \tag{4.180}$$

In this section the following estimation problems are treated: Find gain matrices $G_*^{(i)} \in U_i$, i=1,2, such that

$$\ell_i(G_*^{(i)}) \leq \ell_i(G) \text{ for any } G \in U_i \ , \ i=1,2. \tag{4.181}$$

For each i=1,2, a gain matrix $G_*^{(i)} \in U_i$ for which (4.181) is satisfied, will here be called a *weak optimal gain matrix*.

4.7.2 Sufficient Conditions on Weak Optimal Gain Matrices

4.7.2.1 Case 1

Let $G \in U_1$. Assume that $\zeta_\alpha^G = \{\zeta_\alpha^G(t),\ t \geq 0\}$ is the solution to (4.170).

From the boundedness and continuity of σ and from the definition of $\tilde{U}$ it follows that

$$\{(\int_0^t \sigma_1(X_{s-})dW_1(s),\ldots,\int_0^t \sigma_m(X_{s-})dW_m(s)),\ t \geq 0\} \tag{4.182}$$

and

$$\{(\int_0^t \sum_{j=1}^p g_{1j}(Y_{s-},Z_{s-}^G)dM_j(s),\ldots,\int_0^t \sum_{j=1}^p g_{mj}(Y_{s-},Z_{s-}^G)dM_j(s)),t \geq 0\} \tag{4.183}$$

are (F_t,P) local square integrable martingales and consequently that X_t,Y_t and $Z_t=Z_t^G$ are (F_t,P) local semimartingales (see Boel et al. [107] for the definition of semimartingales).

Let $V : \mathbb{R}^{2m+p} \to \mathbb{R}$ be a bounded and twice differentiable function. Then the differentiation formula (Doléans-Dade and Meyer [109]) yields

$$\begin{aligned}
V(\zeta_\alpha^G(t)) = V(\alpha) &+ \int_0^t \sum_{i=1}^m \frac{\partial V(\zeta_\alpha^G(s-))}{\partial x_i}[f_i(X_{s-})ds + \sigma_i(X_{s-})dW_i(s)] \\
&+ \int_0^t \sum_{i=1}^p \frac{\partial V(\zeta_\alpha^G(s-))}{\partial y_i} c_0[\lambda_i(X_{s-})ds + dM_i(s)] \\
&+ \int_0^t \sum_{i=1}^m \frac{\partial V(\zeta_\alpha^G(s-))}{\partial z_i}\{[f_i(Z_{s-})+c_0 \sum_{j=1}^p g_{ij}(Y_{s-},Z_{s-})(\lambda_j(X_{s-})-\lambda_j(Z_{s-}))]ds \\
&+ c_0 \sum_{j=1}^p g_{ij}(Y_{s-},Z_{s-})dM_j(s)\} + \tfrac{1}{2}\int_0^t \sum_{i=1}^m \sigma_i^2(X_{s-}) \frac{\partial^2 V(\zeta_\alpha^G(s-))}{\partial x_i^2}ds \\
&+ \sum_{s \leq t} [V(\zeta_\alpha^G(s))-V(\zeta_\alpha^G(s-)) - \sum_{i=1}^p \frac{\partial V(\zeta_\alpha^G(s-))}{\partial y_i}(y_i(s)-y_i(s-)) \\
&- \sum_{i=1}^m \frac{\partial V(\zeta_\alpha^G(s-))}{\partial z_i} (z_i(s) - z_i(s-))],
\end{aligned} \tag{4.184}$$

where in the last term the summation is taken over all points of discon= tinuity of ζ_α^G. Recall that if $y_i(t)$ jumps at t=s, then $y_i(s)-y_i(s-)=c_o$ and $z_i(s)-z_i(s-)=c_o\, g_{ii}(Y_{s-},Z_{s-})$. (It is assumed here that $\Delta N_i(s)\Delta N_j(s)=\delta_{ij}$ a.s. for all $s \geq 0$, $i,j=1,\dots,p$, where $\Delta N_i(s)=N_i(s)-N_i(s-)$.).

Hence

$$\sum_{s \leq t} [V(\zeta_\alpha^G(s)) - V(\zeta_\alpha^G(s-)) - \sum_{i=1}^{p} \frac{\partial V(\zeta_\alpha^G(s-))}{\partial y_i}(y_i(s) - y_i(s-)) - \sum_{i=1}^{m} \frac{\partial V(\zeta_\alpha^G(s-))}{\partial z_i}(z_i(s) - z_i(s-))]$$

$$= \sum_{i=1}^{p} \int_0^t [V(X_{s-},Y_{s-}+c_o e^i, Z_{s-}+c_o g_{ii}(Y_{s-},Z_{s-})e^i)-V(\zeta_\alpha^G(s-))]dN_i(s)$$

$$- \sum_{i=1}^{p} \int_0^t \frac{\partial V(\zeta_\alpha^G(s-))}{\partial y_i} c_o dN_i(s) - c_o \sum_{i=1}^{m} \frac{\partial V(\zeta_\alpha^G(s-))}{\partial z_i} \sum_{j=1}^{p} g_{ij}(Y_{s-},Z_{s-})dN_j(s), \tag{4.185}$$

where e^i is the unit vector along the i-th axis in $\mathbb{R}^p$. After some simple manioulations eqns (4.184)-(4.185) yield

$$V(\zeta_\alpha^G(t)) = V(\alpha) + \int_0^t \Big(\sum_{i=1}^{m} \frac{\partial V(\zeta_\alpha^G(s-))}{\partial x_i} f_i(X_{s-})$$

$$+ \sum_{i=1}^{m} \frac{\partial V(\zeta_\alpha^G(s-))}{\partial z_i} [f_i(Z_{s-}) - c_o \sum_{j=1}^{p} g_{ij}(Y_{s-},Z_{s-})\lambda_j(Z_{s-})]$$

$$+ \tfrac{1}{2} \sum_{i=1}^{m} \sigma_i^2(X_{s-}) \frac{\partial^2 V(\zeta_\alpha^G(s-))}{\partial x_i^2} \tag{4.186}$$

$$+ \sum_{i=1}^{p} [V(X_{s-},Y_{s-}+c_o e^i,Z_{s-}+c_o g_{ii}(Y_{s-},Z_{s-})e^i)-V(\zeta_\alpha^G(s-))]\lambda_i(X_{s-})\Big)ds$$

$$+ \Psi(t) \quad , \quad t \geq 0$$

where

$$\Psi(t) \triangleq \int_0^t \Big(\sum_{i=1}^{m} \frac{\partial V(\zeta_\alpha^G(s-))}{\partial x_i} \sigma_i(X_{s-})dW_i(s) \tag{4.187}$$

$$+ \sum_{i=1}^{p} [V(X_{s-},Y_{s-}+c_o e^i,Z_{s-}+c_o g_{ii}(Y_{s-},Z_{s-})e^i)-V(\zeta_\alpha^G(s-))]dM_i(s)\Big), t \geq 0.$$

It can be shown that $\{\Psi(t),\ t \geq 0\}$ is an (F_t,P) local square inte= grable martingale.

Denote

$$\mathcal{L}_1(G)V(\alpha) \triangleq \sum_{i=1}^{m} \frac{\partial V(\alpha)}{\partial x_i} f_i(x) + \sum_{i=1}^{m} \frac{\partial V(\alpha)}{\partial z_i} [f_i(z) - c_0 \sum_{j=1}^{p} g_{ij}(y,z)\lambda_j(z)]$$

$$+ \tfrac{1}{2} \sum_{i=1}^{m} \sigma_i^2(x) \frac{\partial^2 V(\alpha)}{\partial x_i^2} \tag{4.188}$$

$$+ \sum_{i=1}^{p} [V(x,y + c_0 e^i,\ z + c_0 g_{ii}(y,z)e^i) - V(\alpha)]\lambda_i(x).$$

Then eqn (4.186) can be written as

$$V(\zeta_\alpha^G(t)) = V(\alpha) + \int_0^t \mathcal{L}_1(G)V(\zeta_\alpha^G(s-))ds + \Psi(t),\ t \geq 0. \tag{4.189}$$

Let $\mathcal{D}_1$ denote the class of all functions $V = V(\alpha)$ such that: V is continuous on the closure $\bar{D}$ of D and twice continuously differentiable on D; for any $G \in U_1$, $\mathcal{L}_1(G)V \in L_2(D)$. Then using the martingale property of $\{\Psi(t),\ t \geq 0\}$ it follows that

$$E_\alpha V(\zeta_\alpha^G(\tau_1(\alpha;G))) = V(\alpha) + E_\alpha \int_0^{\tau_1(\alpha;G)} \mathcal{L}_1(G)V(\zeta_\alpha^G(s-))ds,\ G \in U,\ V \in \mathcal{D}_1. \tag{4.190}$$

Denote by A the following set in $\mathbb{R}^{2m+p}$:

$$A \triangleq \{\alpha : \alpha \in D \text{ and } |x-z| \leq \varepsilon\} \tag{4.191}$$

and let I_A denote the indicator function of A. Then

$$\Lambda\{t : 0 \leq t < \tau_i(\alpha;\ G),\ |X_t - Z_t^G| \leq \varepsilon\} = \int_0^{\tau_i(\alpha;G)} I_A(\xi(t))dt \tag{4.192}$$

where $\xi(t) = \zeta_\alpha^G(t) = (X_t, Y_t, Z_t^G)$ if i=1, and $\xi(t) = \chi_\alpha^G(t) = (X_t, Y_t, Z_t^G)$ if i=2.

The following lemma will be used in the sequel.

Lemma 4.1

Given $G \in U_1$, Let $V \in \mathcal{D}_1$ satisfy

$$\mathcal{L}_1(G)V(\alpha) = -I_A(\alpha) \ , \ \alpha \in D \tag{4.193}$$

$$V(\alpha) = 0 \ , \quad \alpha \in D^c; \tag{4.194}$$

then

$$V(\alpha) = V_1(\alpha;G) = E_\alpha \Lambda\{t : 0 \le t < \tau_1(\alpha;G), \ |X_t - Z_t^G| \le \varepsilon\} \ . \tag{4.195}$$

Proof

The proof, using eqns (4.190)-(4.192), is straightforward. □

Define the following set of gain matrices

$$U_{01} \triangleq \{G \in U_1 \ : \ V_1(\cdot;G) \in \mathcal{D}_1 \text{ and satisfies eqns (4.193)-(4.194)}\}, \tag{4.196}$$

We are interested here, as in Section 2.6, in the minimization of $\ell_1(G)$ on U_{01} only, $U_{01} \subset U_1$. Define the operator $\mathcal{L}_1^*$

$$\begin{aligned}\mathcal{L}_1^* Q(\alpha) \triangleq -& \sum_{i=1}^{m} \{\partial[f_i(x)Q(\alpha)]/\partial x_i + \partial[f_i(z)Q(\alpha)]/\partial z_i\} \\ &+ (\tfrac{1}{2}) \sum_{i=1}^{m} \partial^2[\sigma_i^2(x)Q(\alpha)]/\partial x_i^2 - Q(\alpha) \sum_{j=1}^{p} \lambda_i(x) \ , \ \alpha \in D\end{aligned} \tag{4.197}$$

for any Q such that $\mathcal{L}_1^* Q \in L_2(D)$.

Applying the same procedure as in the proof of Theorem 2.3, the fol= lowing theorem is obtained.

Theorem 4.4

Suppose $V_{01} \in \mathcal{D}_1$, $G_*^{(1)} \in U_{01}$, and Q_1 satisfy

$$\begin{aligned}&\mathcal{L}_1(G_*^{(1)})V_{01}(\alpha) = -I_A(\alpha), \ \alpha \in D; \quad V_{01}(\alpha) = 0 \ , \ \alpha \notin D \\ &\mathcal{L}_1^* Q_1(\alpha) = T - V_{01}(\alpha) \text{ a.e. in } D; \quad Q_1(\alpha) = 0 \ , \ \alpha \notin D\end{aligned} \tag{4.198}$$

where $G_*^{(1)} \in U_{01}$ is determined by

$$G_*^{(1)} = \arg\sup_{G \in U_{01}} \int_D Q_1(\alpha) \sum_{j=1}^{p} \{c_o \sum_{i=1}^{m} (\partial V(\alpha;G)/\partial z_i) g_{ij}(y,z)\lambda_i(z)$$
$$- V(x,y + c_o e^j,\ z+c_o g_{jj}(y,z)e^j;G)\lambda_j(x)\}d\alpha \ ; \tag{4.199}$$

and $V(\cdot;G)$, $G \in U_{01}$, is the solution to (4.193)-(4.194).

Then

$$\ell_1(G_*^{(1)}) \le \ell_1(G) \text{ for any } G \in U_{01}. \tag{4.200}$$

In this case, the algorithm for computing weak suboptimal gain matrices is the same as described in Section 4.3, except for step 5 which here as= sumes the following form:

5. Compute $G^{(n+1)} = \{g_{ij}^{(n+1)}(y,z) : i=1,\dots,m,\ j=1,\dots,p,\ (y,z) \in D_{yz}\}$ by

$$G^{(n+1)} = \arg\sup_{G \in U_{01}} \int_D Q(\alpha;G^{(n)}) \sum_{j=1}^{p} \{c_o \sum_{i=1}^{m} (\partial V(\alpha;G^{(n)})/\partial z_i) g_{ij}(y,z)\lambda_i(z)$$
$$- V(x,y + c_o e^j,\ z + c_o g_{jj}^{(n)}(y,z)e^j;G^{(n)})\lambda_j(x)\}d\alpha \tag{4.201}$$

where

$$D_x = \{x : |x_i| < 1,\ i=1,\dots,m\},\ D_{yz} = \{(y,z) : |y_j| < 1,\ |z_i| < 1,$$
$$j=1,\dots,p,\ \ i=1,\dots,m\} \tag{4.202}$$

4.7.2.2 Case 2

Let $G \in U_2$. Assume that $\{\chi_\alpha^G(t),\ t \ge 0\}$ is a solution to (4.171). Then, by using martingale theory (and particularly martingale measure theory, Gihman and Skorohod [2]) in a manner similar to that used in Sub= section 4.7.2.1, the following equation is obtained:

$$E_\alpha V(\chi_\alpha^G(\tau_2(\alpha;G))) = V(\alpha) + E_\alpha \int_0^{\tau_2(\alpha;G)} \mathcal{L}_2(G)V(\chi_\alpha^G(s-))ds, \; G \in U_2, \; V \in \mathcal{D}_2, \tag{4.203}$$

where

$$\mathcal{L}_2(G)V(\alpha) \triangleq \sum_{i=1}^{m} f_i(x) \frac{\partial V(\alpha)}{\partial x_i} + \sum_{i=1}^{m} [f_i(z) - \sum_{j=1}^{p} g_{ij}(y,z)c_j(z)] \frac{\partial V(\alpha)}{\partial z_i}$$

$$+ \tfrac{1}{2} \sum_{i=1}^{m} \sigma_i^2(x) \frac{\partial^2 V(\alpha)}{\partial x_i^2} \tag{4.204}$$

$$+ \int_{\mathbb{R}^m} [V(x,y+c(x,u),z + G(y,z)c(x,u))-V(\alpha)]\pi(du), \; \alpha \in D,$$

$$c(x,u) \triangleq (c_1(x,u),\dots,c_p(x,u)), \quad x,u \in \mathbb{R}^m \;, \tag{4.205}$$

and $\mathcal{D}_2$ denotes the class of all functions $V = V(\alpha)$ such that : V is con= tinuous on $\bar{D}$ and twice continuously differentiable on D; for any $G \in U$, $\mathcal{L}_2(G)V \in L_2(D)$.

As in Case 1, the following lemma will be used

Lemma 4.2

Given $G \in U_2$. Let $V \in \mathcal{D}_2$ satisfy

$$\mathcal{L}_2(G)V(\alpha) = -I_A(\alpha) \;, \; \alpha \in D \tag{4.206}$$

$$V(\alpha) = 0 \;, \; \alpha \in D^c \;, \tag{4.207}$$

then

$$V(\alpha) = V_2(\alpha;G) = E_\alpha \Lambda\{t : 0 \le t < \tau_2(\alpha;G), |X_t - Z_t^G| \le \varepsilon\} \;. \tag{4.208}$$

Proof

The proof, using equations (4.203) and (4.192), is straightforward. □

Define the following operator:

$$\mathcal{L}_2^* Q(\alpha) \triangleq - \sum_{i=1}^{m} \{\partial[f_i(x)Q(\alpha)]/\partial x_i + \partial[f_i(z)Q(\alpha)]/\partial z_i\}$$
$$+ (\tfrac{1}{2}) \sum_{i=1}^{m} \partial^2[\sigma_i^2(x)Q(\alpha)]/\partial x_i^2 - Q(\alpha)\pi(\mathbb{R}^m), \ \alpha \in D, \tag{4.209}$$

for any Q such that $\mathcal{L}_2^* Q \in L_2(D)$;

and the following set

$$U_{02} \triangleq \{G \in U_2 : V(\cdot;G) \in \mathcal{D}_2 \text{ and satisfies eqns (4.206)-(4.207)}\}. \tag{4.210}$$

Applying the same procedure as in the proof of Theorem 2.3, the fol= lowing theorem is obtained.

Theorem 4.5

Suppose $V_{02} \in \mathcal{D}_2$, $G_*^{(2)} \in U_{02}$, and Q_2 satisfy

$$\mathcal{L}_2(G_*^{(2)})V_{02}(\alpha) = -I_A(\alpha) \ , \ \alpha \in D; \ V_{02}(\alpha) = 0, \ \alpha \notin D \ , \tag{4.211}$$

$$\mathcal{L}_2^* Q_2(\alpha) = T - V_{02}(\alpha) \text{ a.e. in } D; \ \ Q_2(\alpha) = 0, \ \alpha \notin D \ , \tag{4.212}$$

where $G_*^{(2)} \in U_{02}$ is determined by

$$G_*^{(2)} = \arg\sup_{G \in U_{02}} \int_D Q_2(\alpha)\{ \sum_{i=1}^{m} \sum_{j=1}^{p} g_{ij}(y,z)c_j(z)(\partial V(\alpha;G)/\partial z_i)$$
$$- \int_{\mathbb{R}^m} V(x,y+c(x,u),z + G(y,z)c(x,u);G)\pi(du)\}d\alpha \ ; \tag{4.213}$$

and $V(\cdot;G)$, $G \in U_{02}$, is the solution to (4.206)-(4.207).

Then

$$\ell_2(G_*^{(2)}) \leq \ell_2(G) \text{ for any } G \in U_{02}. \tag{4.214}$$

In this case, the algorithm for computing weak suboptimal gain matrices is the same as described in Section 4.3, except for step 5 which here as= sumes the following form:

5. Compute $G^{(n+1)} = \{g_{ij}^{(n+1)}(y,z) : i=1,\dots,m,\ j=1,\dots,p,\ (y,z) \in D_{yz}\}$ by

$$G^{(n+1)} = \arg\sup_{G \in U_{02}} \int_D Q(\alpha;G^{(n)})\{\sum_{i=1}^{m}\sum_{j=1}^{p} g_{ij}(y,z)c_j(z)(\partial V(\alpha;G^{(n)})/\partial z_i)$$
$$- \int_{\mathbb{R}^m} V(x,y+c(x,u),z + G^{(n)}(y,z)c(x,u);G^{(n)})\pi(du)\}d\alpha. \tag{4.215}$$

4.7.3 Stochastic Stability of Weak Suboptimal Gain Matrices

Let $\mathbb{R}_h^{2m+p}$ be a finite-difference grid on $\mathbb{R}^{2m+p}$, with a constant mesh size h along all axes. Denote $D_h \triangleq \mathbb{R}_h^{2m+p} \cap D$. Given $h > 0$. Then, when= ever the algorithm for computing weak suboptimal gain matrices converges on D_h to a unique solution, we denote it by $\{V_i(\cdot;\bar{G}_i^h),Q_i(\cdot;\bar{G}_i^h),\bar{G}_i^h\}$, i=1,2, where i=1 corresponds to case 1 and i=2 corresponds to case 2.

In order to evaluate the performance of the dynamic state estimators

$$\left.\begin{array}{l} dz_i = f_i(z)dt + \sum_{j=1}^{p} \bar{g}_{ij}^{(1)}(y,z)(dy_j - c_o\lambda_j(z)dt), \\ t > 0,\ i=1,\dots,m,\quad (\bar{G}^{(1)})_{ij} = \bar{g}_{ij}^{(1)} \end{array}\right\} \text{in case 1} \tag{4.216}$$

and

$$\left.\begin{array}{l} dz_i = f_i(z)dt + \sum_{j=1}^{p} \bar{g}_{ij}^{(2)}(y,z)(dy_j-c_j(z)dt) \\ t > 0,\ i=1,\dots,m \quad (\bar{G}^{(2)})_{ij} = \bar{g}_{ij}^{(2)} \end{array}\right\} \text{in case 2} \tag{4.217}$$

where $\bar{G}^{(i)}$ is obtained from $\bar{G}_i^h$ by interpolation and such that $\bar{G}^{(i)} \in U_{oi}$ i=1,2, the following problems have been numerically solved on D_h:

$$\mathcal{L}_i(\bar{G}^{(i)})T_i(\alpha) = -1,\ \alpha \in D;\ T_i(\alpha) = 0,\ \alpha \notin D\ ;\ i=1,2, \tag{4.218}$$

from which it follows that

$$T_i(\alpha) = E_\alpha \tau_i(\alpha;\bar{G}^{(i)}),\ \alpha \in D,\ \ i=1,2. \tag{4.219}$$

The functions T_i, i=1,2, here constitute a measure of the stochastic stability of the system given by (4.170) or (4.171) with $G = \bar{G}^{(i)}$, i=1,2, respectively.

4.8 NUMERICAL EXAMPLES

The following examples are taken from Yavin [79].

4.8.1 Example 1 (Case 1)

Consider the one-dimensional system

$$dx = -\ a_o x dt + \sigma dW,\ t > 0 \tag{4.220}$$

with the measurement process $\{Y_t,\ t \geq 0\}$ determined by

$$dy = c_o dN\ ,\ t > 0\ ,\ y \in \mathbb{R} \tag{4.221}$$

where a_o, σ and c_o are given positive numbers. $\{N(t),\ t \geq 0\}$ is a counting process satisfying $N(t) < \infty$ W.P.1 for all $t \geq 0$. On a given probability space (Ω,F,P) it is assumed that: $\{W(t),\ t \geq 0\}$ is an (F_t,P) Wiener process; and that for a given continuously differentiable function λ, $\{M(t),\ t \geq 0\}$ is a local (F_t,P) square integrable martingale, where

$$M(t) = N(t) - \int_0^t \lambda(X_s)ds,\ t \geq 0. \tag{4.222}$$

Here $F_t = \sigma(X_s, N(s),\ s \leq t)$, and $F_t \subset F$ for all $t \geq 0$. In this case the dynamic state estimator is given by

$$dz = -a_o z dt + g(y,z)(dy - c_o\lambda(z)dt),\ t > 0, z \in \mathbb{R}. \tag{4.223}$$

Thus eqns (4.165) yield

$$\begin{cases} dx = -a_o x dt + \sigma dW \\ dy = c_o\lambda(x)dt + c_o dM \\ dz = [-a_o z + c_o g(y,z)(\lambda(x)-\lambda(z))]dt + c_o g(y,z)dM. \end{cases} \tag{4.224}$$

Here the sets D and A are taken to be

$$D \triangleq \{\alpha = (x,y,z) : |x| < 1, |y| < 1, |z| < 1\} \tag{4.225}$$

$$A \triangleq \{\alpha = (x,y,z) : \alpha \in D \text{ and } |x-z| \leq \varepsilon\} . \tag{4.226}$$

The operators $\mathcal{L}_1(G)$, (4.188), and $\mathcal{L}_1^*$, (4.197), here reduce to

$$\mathcal{L}_1(g)V(\alpha) = -a_0 x \partial V(\alpha)/\partial x + [-a_0 z - c_0 g(y,z)\lambda(z)]\partial V(\alpha)/\partial z + (\sigma^2/2)\partial^2 V(\alpha)/\partial x^2$$
$$+ [V(x,y+c_0, z+c_0 g(y,z)) - V(\alpha)]\lambda(x), \; \alpha \in D, \; V \in \mathcal{D}_1 \tag{4.227}$$

and

$$\mathcal{L}_1^* Q(\alpha) = \partial[a_0 x Q(\alpha)]/\partial x + \partial[a_0 z Q(\alpha)]/\partial z + (\sigma^2/2)\partial^2 Q(\alpha)/\partial x^2 - \lambda(x)Q(\alpha) \tag{4.228}$$

$\alpha \in D$, for any Q such that $\mathcal{L}_1^* Q \in L_2(D)$.

The algorithm for computing weak suboptimal gain matrices, where $g^{(n+1)}$ is determined by

$$g^{(n+1)}(y,z) = g_0 \operatorname{sign} \{\int_{-1}^{1} Q(\alpha;g^{(n)})[c_0(\partial V(\alpha;g^{(n)})/\partial z)\lambda(z)$$
$$- (\lambda(x)/g^{(n)}(y,z))V(x,y+c_0,z+c_0 g^{(n)}(y,z))]dx\} \tag{4.229}$$

$(y,z) \in (-1,1) \times (-1,1)$, $n=0,1,\ldots,$ $G^0 = g_0$;

was applied for the following set of parameters: $\lambda(x) = 10^{-4} + 0.1x^2$, $x \in \mathbb{R}$; $a_0 = 0.0, 0.8$; $\sigma^2 = 0.04, 0.08$; $c_0 = 0.5, 2, 5$; $g_0 = 3,5$; $\varepsilon = 10^{-2}$, 0.1, 0.2; T = 250; h = 0.1. Here the notation $\bar{G}_i^h = \bar{g}_i^h$ (see 4.7.3 for the defi= nition of $\bar{G}_i^h$), i=1,2, is being used.

The numerical results indicate that $\bar{g}^h$ remains the same when $\varepsilon = 0.01$ or $\varepsilon = 0.1$ or $\varepsilon = 0.2$, and that consequently $T_1^h(\alpha) = E_\alpha \tau_1(\alpha;\bar{g}_1^h)$, $\alpha \in D_h$, does not change when ε changes. Also, the numerical results indicate that $V_1(\cdot;\bar{g}_1^h)$, T_1^h and $\bar{g}_1^h$ remain unchanged when c_0 varies among the values {0.5,1,2,5}.

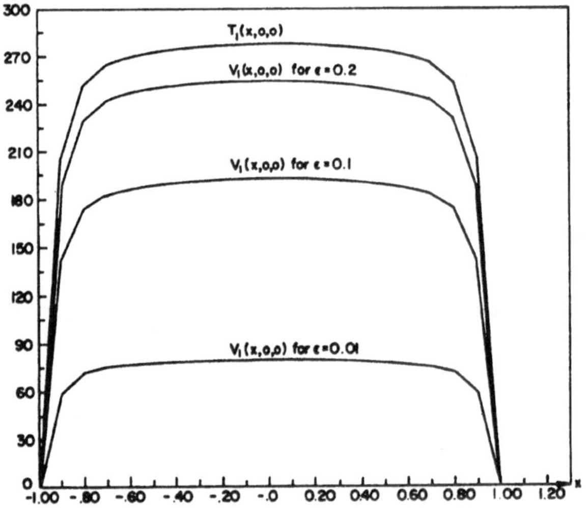

Fig.4.13: The plot of $T_1^h(x,0,0)$ $(T_1(\alpha) = E_\alpha \tau_1(\alpha;\bar{g}_1^h))$ and the plots of $V_1(x,0,0) = V_1(x,0,0;\bar{g}_1^h)$ as functions of x, for $\varepsilon = 0.01$, 0.1, 0.2; $g_0 = 3$, $\sigma^2 = 0.04$ and $a_0 = 0.8$.

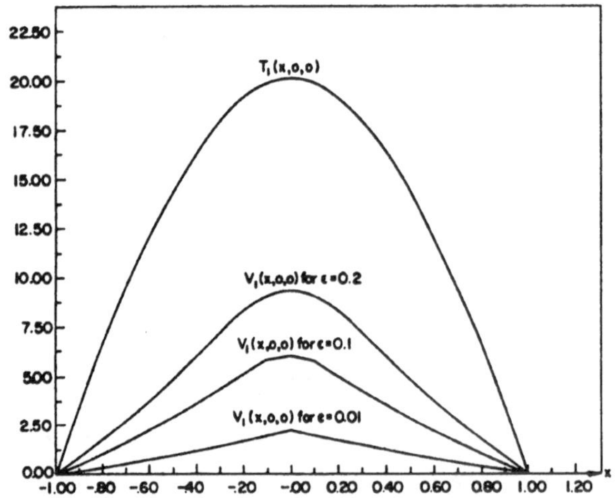

Fig.4.14: The plot of $T_1(x,0,0)(T_1(\alpha) = E_\alpha \tau_1(\alpha;\bar{g}_1^h))$ and the plots of $V_1(x,0,0) = V_1(x,0,0;\bar{g}_1^h)$ as functions of x, for $\varepsilon = 0.01$, 0.1, 0.2; $g_0 = 3$, $\sigma^2 = 0.04$ and $a_0 = 0.0$.

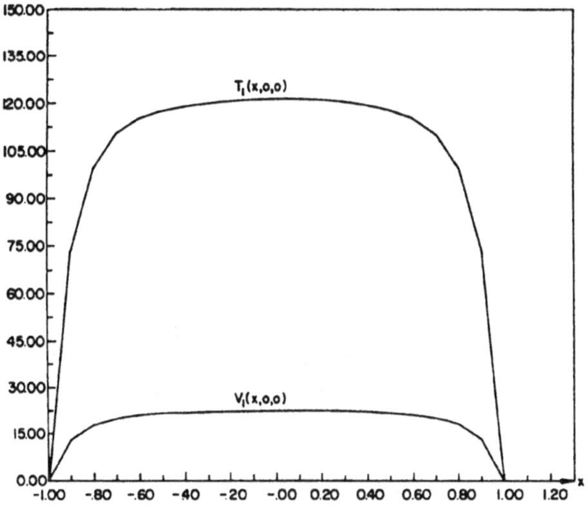

Fig.4.15: The plots of $T_1(x,0,0)(T_1(\alpha) = E_\alpha\tau_1(\alpha;\bar{g}_1^h))$ and $V_1(x,0,0) = V_1(x,0,0;\bar{g}_1^h)$ as functions of x, for $\varepsilon = 0.01$, $g_0 = 3$, $\sigma^2 = 0.08$ and $a_0 = 0.8$.

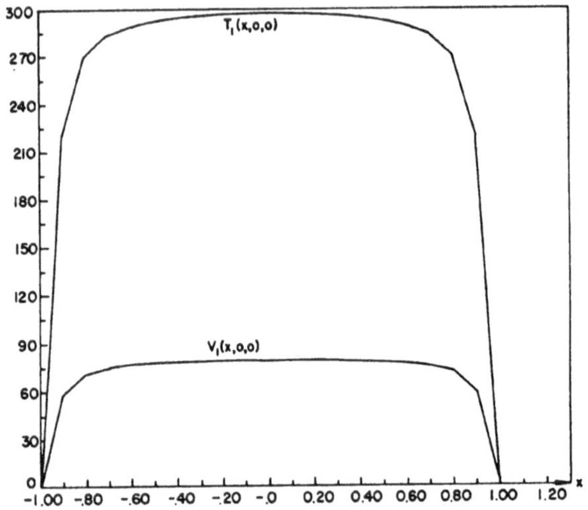

Fig.4.16: The plots of $T_1(x,0,0)(T_1(\alpha) = E_\alpha\tau_1(\alpha;\bar{g}_1^h))$ and $V_1(x,0,0) = V_1(x,0,0;\bar{g}_1^h)$, as functions of x, for $\varepsilon = 0.01$, $g_o = 5$, $\sigma^2 = 0.04$ and $a_o = 0.8$.

4.8.2 Example 2 (Case 2)

Consider the one-dimensional system

$$dx = -a_o x dt + \sigma dW \ , \quad t > 0, \tag{4.230}$$

with the measurement process $\{Y_t, t \geq 0\}$ determined by

$$dy = \int_{\mathbb{R}^m} c(x,u)\nu(dt,du) \ , \ t > 0 \tag{4.231}$$

where $\{\nu(t,A), t \geq 0\}$, $A \in \mathcal{B}(\mathbb{R})$ is a Poisson measure with

$$E\nu(t,A) = t\pi(A) \ , \ t \geq 0, \quad A \in \mathcal{B}(\mathbb{R}) \tag{4.232}$$

and

$$\pi(A) = \begin{cases} 1 & \{1\} \subset A \\ 0 & \{1\} \subset A^c \end{cases} \tag{4.233}$$

Hence $\{Y_t, t \geq 0\}$ can be written as

$$\begin{aligned} Y_t &= \int_0^t c(X_{s-},1)\nu(ds, \{1\}) \\ &= \int_0^t c(X_{s-})\nu(ds,\{1\}) \\ &= \sum_{s \leq t} c(X_{s-})I_{\{1\}}(\nu(s,\{1\}) - \nu(s-,\{1\})) \end{aligned} \tag{4.234}$$

where c(x) is defined in (4.163).

In this case the dynamic state estimator is given by

$$dz = -a_0 z dt + g(y,z)(dy - c(z)dt) \text{ , } t > 0 \text{ ,} \tag{4.235}$$

and eqns (4.167) yield

$$\begin{cases} dx = -a_0 x dt + \sigma dW \\ dy = c(x)dt + c(x)q(dt,\{1\}) \\ dz = [-a_0 z + g(y,z)(c(x) - c(z))]dt + g(y,z)c(x)q(dt,\{1\}). \end{cases} \tag{4.236}$$

The sets D and A are here given by (4.225) and (4.226) respectively.

The operators $\mathcal{L}_2(G)$, (4.204), and $\mathcal{L}_2^*$, (4.209), here reduce to

$$\begin{aligned} \mathcal{L}_2(g)V(\alpha) &= -a_0 x \partial V(\alpha)/\partial x + [-a_0 z - g(y,z)c(z)]\partial V(\alpha)/\partial z + (\sigma^2/2)\partial^2 V(\alpha)/\partial x^2 \\ &\quad + V(x,y+c(x),z + g(y,z)c(x)) - V(\alpha), \ \alpha \in D \text{ , } V \in \mathcal{D}_2 \end{aligned} \tag{4.237}$$

and

$$\mathcal{L}_2^* Q(\alpha) = \partial[a_o x Q(\alpha)]/\partial x + \partial[a_o z Q(\alpha)]/\partial z + (\sigma^2/2)\partial^2 Q(\alpha)/\partial x^2 - Q(\alpha) \tag{4.238}$$

$\alpha \in D$, for any Q such that $\mathcal{L}_2^* Q \in L_2(D)$.

The algorithm for computing weak suboptimal gain matrices, where $g^{(n+1)}$ is determined by

$$g^{(n+1)}(y,z) = g_o \mathrm{sign}\{\int_{-1}^{1} Q(\alpha;g^{(n)})[c(z)\partial V(\alpha;g^{(n)})/\partial z$$

$$- (1/g^{(n)}(y,z))V(x,y+c(x),z + g^{(n)}(y,z)c(x))]dx\} \tag{4.239}$$

$(y,z) \in (-1,1) \times (-1,1)$, $n=0,1,2,\ldots$, $G^0 = g_o$,

was applied for the following set of parameters: $c(x) = 10^{-4} + 0.1x^2$, $x \in \mathbb{R}$; $a_o = 0.4,0.5,0.6,0.7,1.0$; $\sigma^2 = 0.04$, $g_o = 3$; $\varepsilon = 0.01$, $T = 200$ and $h = 0.1$.

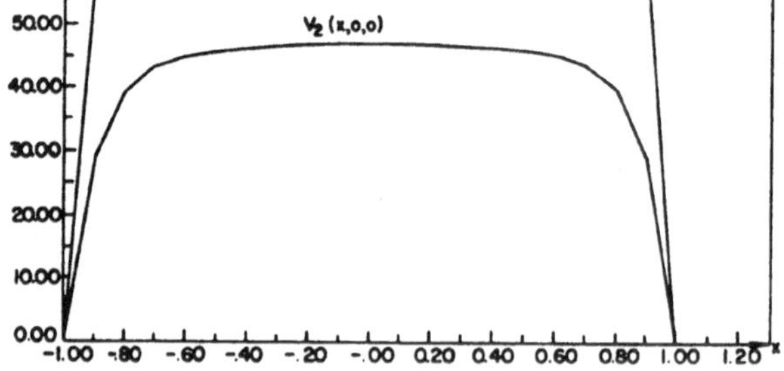

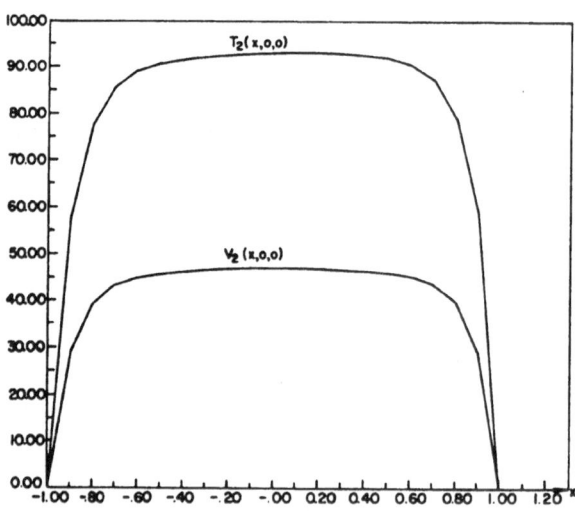

Fig.4.17: The plots of $T_2(x,0,0)(T_2(\alpha) = E_\alpha \tau_2(\alpha;\bar{g}_2^h))$ and $V_2(x,0,0) = V_2(x,0,0;\bar{g}_2^h)$, as functions of x, for $a_o = 0.4$.

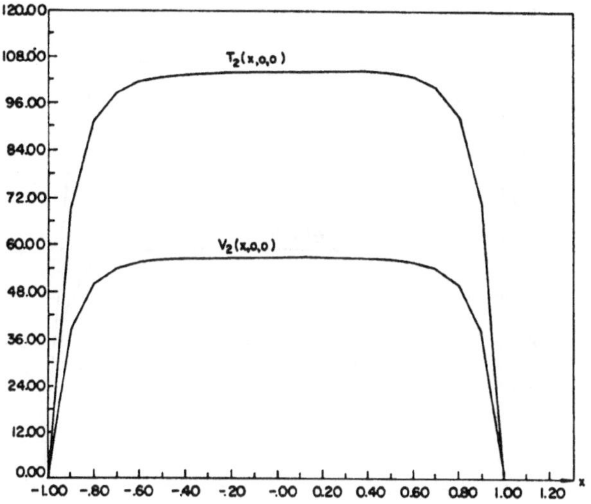

Fig.4.18: The plots of $T_2(x,0,0)(T_2(\alpha) = E_\alpha \tau_2(\alpha;\bar{g}_2^h))$ and $V_2(x,0,0) = V_2(x,0,0;\bar{g}_2^h)$, as functions of x, for $a_o = 0.5$.

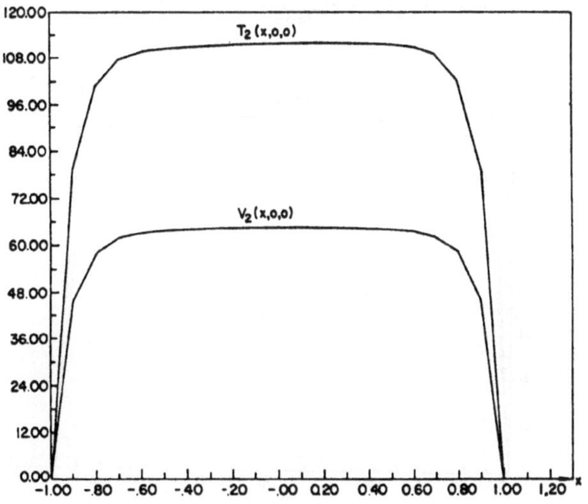

Fig.4.19: The plots of $T_2(x,0,0)(T_2(\alpha) = E_\alpha \tau_2(\alpha;\bar{g}_2^h))$ and $V_2(x,0,0) = V_2(x,0,0;\bar{g}_2^h)$, as functions of x, for $a_o = 0.6$.

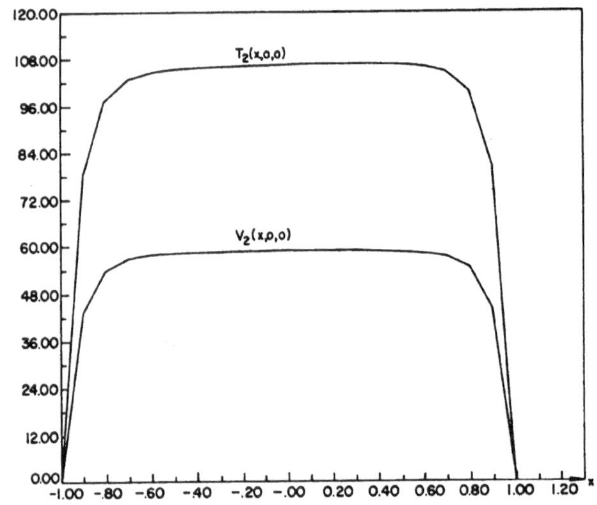

Fig.4.20: The plots of $T_2(x,0,0)(T_2(\alpha) = E_\alpha \tau_2(\alpha;\bar{g}_2^h))$ and $V_2(x,0,0) = V_2(x,0,0;\bar{g}_2^h)$, as functions of x, for $a_o = 0.7$.

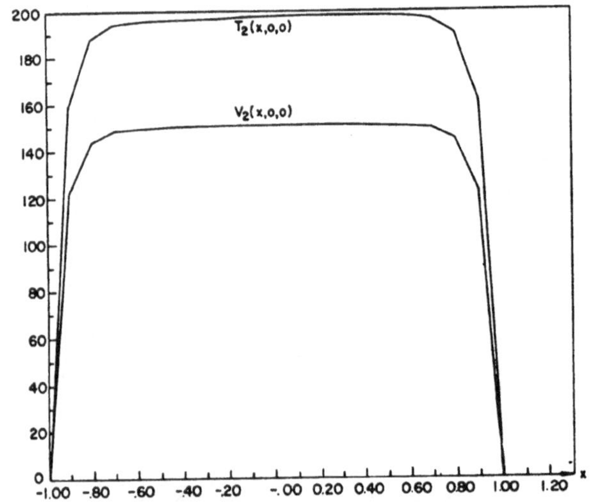

Fig.4.21: The plots of $T_2(x,0,0)(T_2(\alpha) = E_\alpha \tau_2(\alpha;\bar{g}_2^h))$ and $V_2(x,0,0) = V_2(x,0,0;\bar{g}_2^h)$, as functions of x, for $a_o = 1.0$.

REFERENCES

[1] Gihman, I.I., and Skorohod, A.V., *Stochastic Differential Equations*, Springer-Verlag, Berlin, 1972.

[2] Gihman, I.I., and Skorohod, A.V., *The Theory of Stochastic Processes*, Part III, Springer-Verlag, Berlin, 1979.

[3] Segall, A., A Martingale Approach to Modelling, Estimation and Detection of Jump Processes, Technical Report No. 7050-21, Information Systems Laboratory, Stanford University, 1973.

[4] Snyder, D.L., *Random Point Processes*, John Wiley and Sons, New York, 1975.

[5] Fisher, J.R., Optimal nonlinear filtering, in *Advances in Control Systems*, Ed. C.T. Leondes, pp. 197-300, Academic Press, New York, 1967.

[6] Kwakernaak, H., Filtering for systems excited by Poisson white noise; in *Control Theory, Numerical Methods and Computer Systems Modelling*, Edited by A. Bensoussan and J.L. Lions, Int.Symposium, Rocquencourt, June 17-21, 1974.

[7] Tarn, T.J., and Rasis, Y., Observers for nonlinear stochastic systems, *IEEE on Automatic Control*, 21, pp. 441-448, 1976.

[8] Gihman, I.I., and Skorohod, A.V., *Introduction to the Theory of Random Processes*, W.B. Saunders Company, Philadelphia, 1969.

[9] Fleming, W.H., Optimal continuous-parameter stochastic control, *SIAM Review*, 11, pp. 470-509, 1969.

[10] Wonham, W.M., Random differential equations in control theory , in *Probabilistic Methods in Applied Mathematics*, Ed. A.T. Bharucha-Reid, 2, pp. 131-212, Academic Press, New York, 1970.

[11] Fleming, W.H., and Rishel, R.W., *Deterministic and Stochastic Optimal Control*, Springer-Verlag, New York, 1975.

[12] Krylov, N.V., *Controlled Diffusion Processes*, Springer-Verlag, New York, 1980.

[13] Gihman, I.I., and Skorohod, A.V., *Controlled Stochastic Processes*, Springer-Verlag, New York, 1979.

[14] Yavin, Y., and Jordaan, A.M., Optimal controls that maximize the probability of hitting a set of targets - a numerical study, *J.of Optimization Theory and App.*, 34, pp. 517-540, 1981.

[15] Fleming, W.H., Optimal control of partially observable diffu= sions, *SIAM J. Control*, 6, pp. 194-214, 1968.

[16] Ahmed, N.U., and Teo, K.L., An existence theorem on optimal control of partially observable diffusions, *SIAM J.Control*, 12, pp. 351-355, 1974.

[17] Ahmed, N.U., and Teo, K.L., Optimal control of stochastic Ito differential systems with fixed terminal time, *Advances in Applied Probability*, 7, pp. 154-178, 1975.

[18] Reid, D.W., and Teo,K.L., Optimal feedback control of a class of stochastic systems permitting jumps in the diffusion pro= cesses, *Int. J. Systems Science*, 8, pp. 497-511, 1977.

[19] Friedman, M., and Yavin, Y., Optimal control of partially observ= able jump diffusion processes, *Int.J.of Systems Science,* 11, pp. 323-335, 1980.

[20] Davis, M.H.A., and Varaiya, P.P., Dynamic programming conditions for partially observable stochastic systems, *SIAM J. Control,* 11, pp. 226-261, 1973.

[21] Davis, M.H.A., On the existence of optimal policies in stochastic control, *SIAM J. Control,* 11, pp. 587-594, 1973.

[22] Elliott, R.J., and Varaiya, P.P., A sufficient condition for the optimal control of a partially observed stochastic system, Analysis and Optimization of Stochastic Systems (Proc. Inter= national Conf. Univ. Oxford, Oxford, 1978), pp. 11-20, Academic Press, London, 1980.

[23] Christopeit, N., Existence of optimal stochastic controls under partial observation, *Z. Wahrscheinlichkeitstheorie verw. Gebiete,* 51, pp. 201-213, 1980.

[24] Christopeit, N., Optimal stochastic control with special infor= mation patterns, *SIAM J. Control and Optimization,* 18, pp. 559-575, 1980.

[25] Kushner, H.J., On the stochastic maximum principle: fixed time of control, *J. of Math. Analysis and App.*, 11, pp. 78-92, 1965.

[26] Haussmann, U.G., General necessary conditions for optimal control of stochastic systems, *Math. Programming Study,* 6, pp. 30-48, 1976.

[27] Haussmann, U.G., On the stochastic maximum principle, *SIAM J. Control and Optimization*, 16, pp. 236-251, 1978.

[28] Kwakernaak, H., A minimum principle for stochastic control pro= blems with output feedback, *Systems and Control Letters*, 1, pp. 74-77, 1981.

[29] Arkin, V.I., and Saksonov, M.T., Necessary optimality conditions in control problems for stochastic differential equations, *Soviet Math. Dokl.*, 20, pp. 1-5, 1979.

[30] Fleming, W.H., Measure-valued processes in the control of par= tially-observable stochastic systems, *App. Math.Optim.*, 6, pp. 271-285, 1980.

[31] Fleming, W.H., Stochastic control under partial observation, Proc. 4th Int. Conf. Anal & Optimiz. of Systems, INRIA, December 1980.

[32] Ahmed, N.U., *Optimal control of stochastic systems, in Probabilistic Analysis and Related Topics*, (Ed.by A.T.Bharucha-Reid) 2, pp. 1-68, Academic Press, New York, 1979.

[33] Krasovskii, N.N., and Lidskii, E.A., Analytical design of con= trollers with random attributes, I-III, *Automat. Remote Control*, 22, pp. 1021-1025, 1141-1146, 1289-1294, 1961.

[34] Rishel, R., Dynamic programming and minimum principles for sys= tems with jump Markov disturbances, *SIAM J. Control*, 13, pp 338-371, 1975.

[35] Sworder, D.D., Feedback control of a class of linear systems with jump parameters, *IEEE Trans. Automatic Control*, 14, pp.9-14, 1969.

[36] Olsder, G.J. and Suri, R., Time-optimal control of parts-routing in a manufacturing system with failure-prone machines, Proc. 19th IEEE Conf. on Decision & Control, 1, pp. 722-727, 1980.

[37] Sawaragi, Y., Katayama, T., and Fujishige, S., State estimation for continuous-time system with interrupted observation, *IEEE Trans. on Automatic Control*, 19, pp. 307-314, 1974.

[38] Yavin, Y., and Venter, A., Optimal control of stochastic systems with interrupted observation, *Computers and Mathematics with Applications*, 7, pp. 509-525, 1981.

[39] Yavin, Y., Bang-bang strategies using interrupted observations for steering a random motion of a point, *Computer Methods in Applied Mechanics and Eng.*,29, pp. 351-364, 1981.

[40] Yavin, Y., Strategies using interrupted observations for hitting a moving target, *Int. J. of Systems Sci.*, 13, pp. 159-175,1982.

[41] Kailath, T., A view of three decades of linear filtering theory, *IEEE Trans. on Information Theory*, 20, pp. 146-181, 1974.

[42] Kalman, R., A new approach to linear filtering and prediction problems, *J. Basic Eng.* (Trans. ASME, Series D), 82, pp. 35-45, 1960.

[43] Kalman, R., and Bucy, R., New results in linear filtering and prediction theory, *J. Basic Eng.* (Trans. ASME, Series D), 83, pp. 95-108, 1961.

[44] Stratonovich, R., On the theory of optimal nonlinear filtration of random functions, *Theory of Probability and its Applications*, 4, pp. 223-225, 1959.

[45] Kushner, H.J., On the dynamical equations of conditional proba= bility density functions with applications to optimal stochas= tic control theory, *J. Math. Anal. App.*, 8, pp. 332-344, 1964.

[46] Kushner, H.J., On the differential equations satisfied by con= ditional probability densities of Markov processes, with applications, *J. SIAM Control*, 2, pp. 106-119, 1964.

[47] Wonham, W.M., Some applications of stochastic differential equa= tions to optimal nonlinear filtering, *J. SIAM Control*, 2, pp. 347-369, 1965.

[48] Bucy, R.S., Nonlinear filtering theory, *IEEE Trans. on Automatic Control*, 10, pp. 198,1965.

[49] Bucy, R.S., and Joseph, P.D., *Filtering for Stochastic Processes with Applications to Guidance*, Interscience, New York, 1968.

[50] Jazwinski, A.H., *Stochastic Processes and Filtering Theory*, Academic Press, New York, 1970.

[51] Frost, P.A., and Kailath, T.K., An innovations approach to least-squares estimation - Part III : Nonlinear estimation in white Gaussian noise, *IEEE Trans. Automat. Contr.*, 16, pp.217-226, 1971.

[52] McGarty, T.P., *Stochastic Systems and State Estimation*, John Wiley & Sons, New York, 1974.

[53] Fujisaki, M., Kallianpur, G., and Kunita, H., Stochastic differen= tial equations for the nonlinear filtering problem, *Osaka J.Math.*, 9, pp. 19-40, 1972.

[54] Snyder, D.L., Filtering and detection for doubly stochastic Poisson processes, *IEEE Trans. on Information Theory*, 18, pp. 91-102, 1972.

[55] Clements, D., and Anderson, B.D.O., A nonlinear fixed-lag smoother for finite state Markov process, *IEEE Trans. on Information Theory*, 21, pp. 446-452, 1975.

[56] Liptser, R.S., and Shiryayev, A.N., Statistics of Random Processes, Springer-Verlag, Berlin, Part I : 1977, Part II 1978.

[57] Björk, T., Finite dimensional optimal filters for a class of Itô processes with jumping parameters, *Stochastics*, 4, pp. 167-183, 1980.

[58] Rishel, R., A comment on a dual control problem, *IEEE Trans. on Automat. Contr.*, 26, pp. 606-609, 1981.

[59] Mehra, R., A comparison of several nonlinear filters for reentry vehicle tracking, *IEEE Trans. Automatic Control*, 16, pp. 307-319, 1971.

[60] Athans, M., The role and use of the stochastic linear-quadratic-Gaussian problem in control system design, *IEEE Trans. on Automatic Control*, 16, pp. 529-552, 1971.

[61] McGarty, T.P., The estimation of the constituent densities of the upper atmosphere by means of a recursive filtering algo= rithm, *IEEE Trans. on Automatic Control*, 16, pp. 817-823,1971.

[62] Dressler, R.M., and Tabak, D., Satellite tracking by combined optimal estimation and control techniques, *IEEE Trans. on Automatic Control*, 16, pp. 833-840, 1971.

[63] Snyder, D.L., and Rhodes, I.B., Phase and frequency tracking accuracy in direct-detection optical-communication systems, *IEEE Trans. on Communications*, 20, pp. 1139-1142, 1972.

[64] Davidson, F.M., and Carlson, R.T., Point process estimators of Gaussian optical field intensities, *IEEE Trans. on Information Theory*, 25, pp. 620-624, 1979.

[65] Baras, J.S., Dorsey, A.J., and Levine, W.S., Estimation of traffic platoon structure from headway statistics, *IEEE Trans. on Automatic Control*, 24, pp. 553-559, 1979.

[66] Bagchi, A., and Van Maarseveen, M., Modelling and estimation of traffic flow - a martingale approach, *Int.J. Systems Science*, 11, pp. 429-444, 1980.

[67] Van Schuppen, J.H., Estimation Theory for Continuous Time Pro= cesses, A Martingale Approach, Memorandum No. ERL-M405, Elec= tronic Research Laboratory, College of Eng., University of California, Berkeley, 1973.

[68] Van Schuppen, J.H., Filtering, prediction and smoothing for counting process observations, a martingale approach, *SIAM J.App. Math.*, 32, pp. 552-570, 1977.

[69] Segall, A., and Kailath, T., The modelling of randomly modulated jump processes, *IEEE Trans. on Information Theory*, 21, pp. 135-143, 1975.

[70] Segall, A., Davis, M.H.A., and Kailath, T., Nonlinear filtering with counting observations, *IEEE Trans. on Information Theory*, 21, pp. 143-149, 1975.

[71] Vaca, M.V., and Snyder, D.L., Estimation and decision for obser= vations derived from martingales: Part I, representation, *IEEE Trans. on Information Theory*, 22, pp. 691-707, 1976.

[72] Vaca, M.V., and Snyder, D.L., Estimation and decision for obser= vations derived from martingales : Part II, *IEEE on Information Theory*, 24, pp. 32-45, 1978.

[73] Gertner, I., An alternative approach to nonlinear filtering, *Stochastic Processes and their Applications*, 7, pp. 231-246, 1978.

[74] Boel, R.K., and Beneš, V.E., Recursive nonlinear estimation of a diffusion acting as the rate of an observed Poisson process, *IEEE Trans. on Information Theory*, 26, pp. 561-575, 1980.

[75] Wan, C.B., and Davis, M.H.A., The general point process disorder problem, *IEEE Trans. on Information Theory*, 23, pp. 538-540, 1977.

[76] Davis, M.H.A., and Andreadakis, E., Exact and approximate fil= tering in signal detection: an example, *IEEE Trans. on Infor= mation Theory*, 23, pp. 768-772, 1977.

[77] Yavin, Y., and Friedman, M., Estimation and control for a class of nonlinear stochastic systems, *Int. J. Systems Science*, 12, pp 587-600, 1981.

[78] Yavin, Y., An alternative approach to nonlinear filtering, *Int. J. of Systems Science*, 12, pp. 795-812, 1981.

[79] Yavin, Y., An alternative approach to nonlinear filtering: jump process observations, *Int. J. Systems Science*, 12, pp. 1061-1081, 1981.

[80] Yavin, Y., An alternative approach to nonlinear filtering: maximizing the probability of hitting a target set, *Int.J. of Systems Science*, 13, pp. 289-299, 1982.

[81] Dynkin, E.B., *Markov Processes*, Vol.I, II, Springer-Verlag, Berlin, 1965.

[82] Yosida, K., *Functional Analysis*, Springer-Verlag, New York, 1968.

[83] Friedman, M., and Yavin, Y., Optimal controls that maximize the expectation of first passage time, *J. Franklin Inst.*, 304, pp 231-242, 1977.

[84] Friedman, M., and Yavin, Y., On the numerical solution of two coupled nonlinear partial integro-differential equations related to the optimal control of a nonlinear noisy oscilla= tor, *Computer Methods in Appl. Mechanics and Eng.*, 16, pp. 37-46, 1978.

[85] Yavin, Y., and Friedman, M., On the optimal control of a stochastic system with discontinuous sample paths, *Int.J.Systems Science*, 9, pp. 451-465, 1978.

[86] Stroock, D.W., Diffusion processes associated with Levy genera= tors, *Z.Wahrscheinlichkeitstheorie verw. Gebiete* 32, pp 209-244, 1975.

[87] Mahno, S.Ja., On weak solutions of stochastic differential equa= tions, *Theor. Probability and Math.Statist.*, No.13, pp. 116-124, 1977.

[88] Komatsu, T., Markov processes associated with certain integro-differential operators, *Osaka J. Math.*, 10, pp. 271-303, 1973.

[89] Stroock, D.W., and Varadhan, S.R.S., Diffusion processes with continuous coefficients,I,II, *Communications on Pure and App Math.*, 22, pp 345-400, 479-530, 1969.

[90] Benes, V.E., Existence of optimal stochastic control laws, *SIAM J Control*, 9, pp. 446-472, 1971.

[91] Yavin, Y., and Reuter, G.W., Computation of Nash equilibrium pairs of a stochastic differential game, *Optimal Control Applications and Methods*,2, pp. 225-238, 1981.

[92] Neveu, J., *Mathematical Foundations of the Calculus of Probability*, San Francisco, Holden-Day Inc., 1965.

[93] Kushner, H.J. and DiMasi, G., Approximations for functionals and optimal control problems on jump diffusion processes, *J. of Math. Analys. and App.*, 63, pp. 772-800, 1978.

[94] Kushner, H.J., *Probability Methods for Approximations in Stochastic Control and for Elliptic Equations*, New York, Academic Press, 1977.

[95] Yavin, Y., and Reuter, G.W., Optimal bang-bang control of par= tially observable stochastic systems, *Int.J. Systems Science*, 12, pp. 147-161, 1981.

[96] Yavin,Y., Bang-bang partially observable feedback strategies for a rendezvous problem, *Int. J.Systems Science*, 12, pp.1417-1428,1981.

[97] Prohorov, Y.V., and Rozanov, Y.A., *Probability Theory*, Springer-Verlag, Berlin, 1969.

[98] Strümpfer, J., Search Theory Index, TN-017-80, Institute for Maritime Technology, Simonstown, South Africa, 1980.

[99] Sergeeva, L.V. and Teterina, N.I., Investigation of the solution of a stochastic equation with random coefficients, *Theor.Probability and Math. Statis.*, pp. 145-158, 1974.

[100] Sergeeva, L.V., On a certain generalization of diffusion processes, *Theor. Probability and Math. Statist.*, pp. 161-169, 1976.

[101] Kushner, H.J., *Stochastic Stability and Control*, Academic Press, New York, 1967.

[102] Yavin, Y. and Jordaan, A.M., Optimal controls that maximize the probability of hitting a moving target, *Computers and Math. with Appls.*, 7, pp. 17-26, 1981.

[103] Yavin, Y., Suboptimal strategies for steering a random motion of a point in a multitarget environment, Technical Report, TWISK 250, NRIMS, CSIR, Pretoria, March 1982.

[104] Bar-Shalom, Y., Tracking methods in a multitarget environment, *IEEE Trans. on Automatic Control*, AC-23, pp. 618-626, 1978.

[105] Bar-Shalom, Y., and Marcus, G.E., Tracking with measurements of uncertain origin and random arrival times, *IEEE Trans. on Automatic Control*, AC-25, pp. 802-807, 1980.

[106] Crandel, S.H., and Mark, W.D., *Random Vibrations in Mechanical Systems*, Academic Press, New York, 1963.

[107] Boel, R., Varaiya, P., and Wong, E., Martingales on jump processes: I: Representation results. II: Applications, *SIAM J. Control*, 13, pp. 999-1021, 1022-1061, 1975.

[108] Meyer, P.A., *Probability and Potentials*, Blaisdell, San Francisco, 1966.

[109] Doleans-Dade, C., and Meyer, P.A., *Seminaire de Probabilities IV*, pp. 77-107, Lecture Notes in Mathematics, No.124, Springer-Verlag, New York, 1970.

Lecture Notes in Control and Information Sciences

Edited by A. V. Balakrishnan and M. Thoma

Vol. 22: Optimization Techniques
Proceedings of the 9th IFIP Conference on
Optimization Techniques,
Warsaw, September 4–8, 1979
Part 1
Edited by K. Iracki, K. Malanowski, S. Walukiewicz
XVI, 569 pages. 1980

Vol. 23: Optimization Techniques
Proceedings of the 9th IFIP Conference on
Optimization Techniques,
Warsaw, September 4-8, 1979
Part 2
Edited by K. Iracki, K. Malanowski, S. Walukiewicz
XV, 621 pages. 1980

Vol. 24: Methods and Applications
in Adaptive Control
Proceedings of an International Symposium
Bochum, 1980
Edited by H. Unbehauen
VI, 309 pages. 1980

Vol. 25: Stochastic Differential Systems –
Filtering and Control
Proceedings of the IFIP-WG7/1 Working Conference
Vilnius, Lithuania, USSR, Aug. 28 – Sept. 2, 1978
Edited by B. Grigelionis
X, 362 pages. 1980

Vol. 26: D. L. Iglehart, G. S. Shedler
Regenerative Simulation of Response
Times in Networks of Queues
XII, 204 pages. 1980

Vol. 27: D. H. Jacobson, D. H. Martin, M. Pachter, T. Geveci
Extensions of Linear-Quadratic Control Theory
XI, 288 pages. 1980

Vol. 28: Analysis and Optimization of Systems
Proceedings of the Fourth International
Conference on Analysis and Optimization of Systems
Versailles, December 16–19, 1980
Edited by A. Bensoussan and J. L. Lions
XIV, 999 pages. 1980

Vol. 29: M. Vidyasagar,
Input-Output Analysis of Large-Scale
Interconnected Systems –
Decomposition, Well-Posedness and Stability
VI, 221 pages. 1981

Vol. 30: Optimization and Optimal Control
Proceedings of a Conference Held at
Oberwolfach, March 16–22, 1980
Edited by A. Auslender, W. Oettli, and J. Stoer
VIII, 254 pages. 1981

Vol. 31: Berc Rustem
Projection Methods in Constrained
Optimisation and Applications
to Optimal Policy Decisions
XV, 315 pages. 1981

Vol. 32: Tsuyoshi Matsuo,
Realization Theory of
Continuous-Time Dynamical Systems
VI, 329 pages, 1981

Vol. 33: Peter Dransfield
Hydraulic Control Systems –
Design and Analysis of Their Dynamics
VII, 227 pages, 1981

Vol. 34: H.W. Knobloch
Higher Order Necessary Conditions
in Optimal Control Theory
V, 173 pages, 1981

Vol. 35: Global Modelling
Proceedings of the IFIP-WG 7/1 Working
Conference Dubrovnik, Yugoslavia,
Sept. 1–5, 1980
Edited by S. Krčevinac
VIII, 232 pages, 1981

Vol. 36: Stochastic Differential Systems
Proceedings of the 3rd IFIP-WG 7/1
Working Conference
Visegrád, Hungary, Sept. 15–20, 1980
Edited by M. Arató, D. Vermes, A.V. Balakrishnan
VI, 238 pages, 1981

Vol. 37: Rüdiger Schmidt
Advances in Nonlinear
Parameter Optimization
VI, 159 pages, 1982

Vol. 38: System Modeling and Optimization
Proceedings of the 10th IFIP Conference
New York City, USA, Aug. 31 – Sept. 4, 1981
Edited by R.F. Drenick and F. Kozin
XI, 894 pages. 1982

Vol. 39: Feedback Control of
Linear and Nonlinear Systems
Proceedings of the Joint Workshop
on Feedback and Synthesis of
Linear and Nonlinear Systems
Bielefeld/Rom
XIII, 284 pages. 1982

Vol. 40: Y.S. Hung, A.G.J. MacFarlane
Multivariable Feedback:
A Quasi-Classical Approach
X, 182 pages. 1982

Vol. 41: M. Gössel
Nonlinear Time-Discrete Systems –
A General Approach by
Nonlinear Superposition
VIII, 112 pages. 1982

Vol. 42: Advances in Filtering and
Optimal Stochastic Control
Proceedings of the IFIP-WG 7/1
Working Conference
Cocoyoc, Mexico, February 1–6, 1982
VIII, 391 pages. 1982

Lecture Notes in Control and Information Sciences

Edited by A. V. Balakrishnan and M. Thoma

Vol. 43: Stochastic Differential Systems
Proceedings of the 2nd Bad Honnef Conference
of the SFB 72 of the DFG at the University of Bonn
June 28 – July 2, 1982
Edited by M. Kohlmann and N. Christopeit
XII, 377 pages. 1982.

Vol. 44: Analysis and Optimization of Systems
Proceedings of the Fifth International
Conference on Analysis and Optimization of Systems
Versailles. December 14–17, 1982
Edited by A. Bensoussan and J. L. Lions
XV, 987 pages, 1982

Vol. 45: M. Arató
Linear Stochastic Systems
with Constant Coefficients
A Statistical Approach
IX, 309 pages. 1982

Vol. 46: Time-Scale Modeling of Dynamic Networks
with Applications to Power Systems
Edited by J. H. Chow
X, 218 pages. 1982

Vol. 47: P. A. Ioannou, P. V. Kokotovic
Adaptive Systems with Reduced Models
V, 162 pages. 1983

Vol. 48: Yaakov Yavin
Feedback Strategies for Partially
Observable Stochastic Systems
VI, 233 pages. 1983